The Ethics of Animal Re-creation and Modification

The Ethics of Animal Re-creation and Modification

Reviving, Rewilding, Restoring

Edited by

Markku Oksanen and Helena Siipi

palgrave
macmillan

First published 2014 by
PALGRAVE MACMILLAN

Palgrave Macmillan in the UK is an imprint of Macmillan Publishers Limited, registered in England, company number 785998, of Houndmills, Basingstoke, Hampshire RG21 6XS.

Palgrave Macmillan in the US is a division of St Martin's Press LLC, 175 Fifth Avenue, New York, NY 10010.

Palgrave Macmillan is the global academic imprint of the above companies and has companies and representatives throughout the world.

Palgrave® and Macmillan® are registered trademarks in the United States, the United Kingdom, Europe and other countries.

ISBN 978–1–137–33763–4

This book is printed on paper suitable for recycling and made from fully managed and sustained forest sources. Logging, pulping and manufacturing processes are expected to conform to the environmental regulations of the country of origin.

A catalogue record for this book is available from the British Library.

A catalog record for this book is available from the Library of Congress.

Contents

Acknowledgements

We would like to thank all the people in the philosophy unit at the University of Turku, Finland, and especially the biologist Timo Vuorisalo, who showed interest in this ground-breaking project and helped us to organize a workshop on the theme discussed in this book in December 2011. We thank all the contributors for their participation. We would also like to thank Kreeta Ranki, who helped us to assemble the typescript, and Susanne Uusitalo, who checked the language of our joint and individual contributions. This project has been supported by two Academy of Finland grants (218139 and 121018).

MARKKU OKSANEN
HELENA SIIPI

Notes on the Contributors

Elisa Aaltola is Senior Lecturer in Philosophy in the Department of Social Sciences, University of Eastern Finland. She is the author of *Animal Suffering: Philosophy and Culture* (2012), and has in addition published numerous articles and three books (in Finnish) on animal ethics and more broadly animal philosophy.

Julien Delord is a philosopher and historian of ecology and environmental sciences. After his PhD on the historical, epistemological and ethical aspects of the concept of species extinction, he taught at the École Normale Supérieure, Paris, and at the University of Brest, France. Today, he conducts research on the epistemological and political dimensions of biodiversity conservation and restoration in Toulouse. He is the author of *L'extinction d'espèce: histoire d'un concept et enjeux éthiques* (2010).

Christian Gamborg is Associate Professor in Natural Resource Ethics at the Department of Food and Resource Economics, University of Copenhagen, Denmark. Since 1998 the major part of his research and teaching has concerned environmental and natural resource ethics in relation to topics such as nature restoration, forest and wildlife management, modern biotechnology, breeding, and biofuels. He also teaches philosophy of science in relation to landscape architecture as well as environmental conflict management.

Bart Gremmen is Professor of Ethics in Life Sciences at Wageningen University, the Netherlands. He is also a senior research associate at the Oxford Uehiro Centre for Practical Ethics, University of Oxford, UK. His current research focuses on the ethical and societal issues in emerging technologies, genomics, nanotechnology, genetic engineering and synthetic biology. He has published on a wide range of topics, including sustainability, hermeneutics, waste disposal, discursive psychology, the precautionary principle, the de-domestication of animals, novel foods, theory of practice, rules, pragmatism, celiac disease, allergy and animal welfare.

Anne I. Myhr is Acting Director at GenØk, the Centre for Biosafety in Tromsø, Norway. She holds a Master's in biotechnology from the

Norwegian University of Science and Technology (NTNU), Trondheim, and a PhD from the University of Tromsø. Her research interests include innovation processes and implications of emergent technologies, such as GMOs and nanobiotechnology, and capacity building in risk assessment and management of GMO use and release in developing countries. She has been a member of several national and Nordic scientific committees and has internationally been involved in various issues related to GMOs, including socio-economic impacts under the Cartagena Protocol on Biosafety.

Bjørn K. Myskja is Professor of Ethics and Political Philosophy at the Department of Philosophy and Religious Studies, the Norwegian University of Science and Technology (NTNU), Trondheim. For the academic year 2013–14, he is Guest Professor at the Department of Food and Resource Economics, the University of Copenhagen. His research interests include bioethics, ethics of technology, Kantian ethics, Aristotelian ethics, political philosophy and aesthetics. He has published on subjects of trust, expertise and lay engagement regarding bio- and nanotechnologies, as well as on theoretical ethics. He is a research partner on a number of multidisciplinary, integrated research projects within bio- and nanotechnology, and is a member of several Norwegian national ethics committees and advisory boards.

Markku Oksanen is Academy Research Fellow in the Department of Behavioural Sciences and Philosophy, University of Turku, Finland. He is the co-editor of *Philosophy and Biodiversity* (2004). He has published widely on environmental philosophy, particularly on issues relating to property rights, green political theory, biodiversity and climate change.

Helena Siipi is Lecturer in Philosophy at the University of Turku, Finland. Her research interests include environmental ethics and applied ethics, especially ethics of new biotechnologies and the value of naturalness. She has published numerous articles on these topics.

Derek Turner is Associate Professor of Philosophy and Chair of the Philosophy Department at Connecticut College in New London, CT, USA. He is the author of *Making Prehistory: Historical Science and the Scientific Realism Debate* (2007) as well as *Paleontology: A Philosophical Introduction* (2011). He has also published several articles in environmental philosophy, on topics ranging from the role of metaphor in environmental thinking to the ethics of NIMBY ('not-in-my-backyard') claims.

Introduction: Towards a Philosophy of Resurrection Science

Markku Oksanen and Helena Siipi

Extinction and the idea of de-extinction

Bringing extinct species back to life may be possible in the near future. If researchers are successful, this would challenge the basic tenet of conservation, and extinction need no longer be forever. As revolutionary as the idea of reversibility of extinction may sound, the idea of extinction itself – the possibility of species dying out and disappearing – has in certain periods of history appeared equally world-shattering.

The general view in antiquity was that extinctions had occurred and that fossils were proof of this (Mayor 2011). However, Plato and Aristotle somewhat avoided the topic and this may have influenced the disappearance of the belief in Europe later, when creationist views prevailed and accordingly species were considered eternal and fixed entities. It was believed that long vanished animals might live elsewhere or exist in transmuted forms (Mayr 1988, p. 203; Moore 1999, pp. 107–19) In the late eighteenth and early nineteenth century, the Western belief in the eternalness of species was still so powerful that, as the science historian Martin Rudwick (2005, p. 243) puts it, 'most naturalists felt an almost gut revulsion against the idea that extinction might be an ordinary feature of the natural world'. However, through systematic palaeontological and biological studies this belief started to wane and the pace of change of commonly held beliefs became noticeable within the span of a single human life. For example, in his *Notes on the State of Virginia* (1781), the US President Thomas Jefferson (1743–1826) wrote famously:

> The bones of the Mammoth which have been found in America, are as large as those found in the old world. It may be asked, why I insert

> the Mammoth, as if it still existed? I ask in return, why I should omit it, as if it did not exist? Such is the economy of nature, that no instance can be produced of her having permitted any one race of her animals to become extinct; of her having formed any link in her great work so weak as to be broken. (Jefferson 1999, p. 55)

Jefferson was so convinced of the existence of a mastodon – a species of the genus *Mammut* – that he anticipated the famous Lewis and Clark expedition (1804–06) to find living specimens of the species.[1] However, the appearance of new scientific knowledge made Jefferson revise his beliefs and, in private discussions before his death in 1826, he admitted the possibility of extinction (Barrow 2009, pp. 18–19).

The publication of Charles Darwin's *On the Origin of Species* (1859) is the scientific culmination of the change of beliefs concerning extinction. Darwin established the metaphysical foundations for conservation biology: 'when a group has once wholly disappeared, it does not reappear' (Darwin 1998, p. 277). Darwin himself was not too worried about extinctions. He saw them as a vital part of the evolutionary process and believed that the overall number of species remained relatively unaltered (ibid., pp. 258–9; Cuddington and Ruse 2004). However, we now know that several mass extinctions have taken place on earth. It is for sure that the prehistoric ones are solely due to the forces of nature, though the arrival and triumph of *Homo sapiens* means that more recent extinctions have to be explained differently. These facts and Darwin's idea of irreversibility of extinctions constitute the ethical foundation for conservation biology: the designation of human responsibilities with respect to biodiversity.

So far, our responsibilities towards biodiversity have manifested themselves as prima facie obligations not to destroy species. Now, with new knowledge, and through emerging biotechnological possibilities, it may well be asked whether we also have a duty to repair the species destruction caused by humans and so 'restore the balance of justice between humans and nonhumans' (Taylor 1986, p. 305). In many projects all over the world, researchers are working towards an ambitious goal of re-creating different extinct species from the passenger-pigeon to the thylacine. As a case in point, the resurrection of the woolly mammoth is currently (as of August 2013) the target of two competing research groups; and the renowned palaeontologist Jack Horner actually ponders 'reverse evolution' and whether we might build a dinosaur from a chicken (Horner and Gorman 2010). Along with these plans, new linguistic expressions have been coined. To reverse the extinction of a species is to make it 'de-extinct'; the process is termed 'de-extinction'. Terms such as 're-creation', 'resurrection', 'reviving' and 'resuscitation' are commonly used to refer to this action.

The new biotechnology and scientific knowledge offers new possibilities, not just for reversing extinctions but also for avoiding them. Even though some anthropogenic extinctions result from hunting, most follow indirectly from habitat fragmentation and the introduction of invasive alien species. The shortcomings in traditional conservation may increase the temptation to apply gene-technological means. Among these possibilities of gene-technological conservation are the applications of synthetic biology and different forms of biocontrol, as discussed by Anne I. Myhr and Bjørn K. Myskja in Chapter 6. Biocontrol refers to the techniques in which harmful pests are either genetically modified to be less harmful or incapacitated by genetically modified viruses. Another possibility is to modify genetically the endangered animal species and thus help them to adapt to their changing (or new) environment. According to a radical idea, tigers and other endangered wild animals could be made 'domestic pets', as discussed by Elisa Aaltola in Chapter 5.

As the idea of extinction caused 'an almost gut reaction' to eighteenth-century naturalists, so do the ideas of de-extinction and the gene-technological conservation of endangered species seem to affect many conservation biologists (see for example Rogers 2002, pp. 158–9; Pimm 2013), the advocates of animal rights, and members of the general public in a similar way. Why are these ideas so contentious? Popular books, films and documentaries on giant monsters, such as *King Kong*, *Jurassic Park* and *Walking with Dinosaurs*, provide the most obvious answer: velociraptors, sabre-toothed tigers and the like are dangerous animals and difficult, if not impossible, to keep under control. As topical as these answers are still in the world of persistent conflicts between man-eaters and human beings, between the predators and the prey, this only applies to a very small minority of cases of de-extinction and gene-technological conservation; but many in the re-creation field want to go beyond this. Many of the candidate species for resurrection and genetic modification are as harmless as average sized birds and butterflies.

But are all average sized birds and butterflies harmless – especially if genetically modified? Do the ideas of de-extinction and gene-technological conservation rest on a philosophically solid basis? The collection of essays in this book aims to assess the biological and philosophical underpinnings of these possibilities critically. The critical approach focuses on, in particular, the nature and the value of de-extinct and genetically modified animals, the moral costs of the development of the methods, and their application to real-life situations in conservation. We shall specify these issues further in this introduction, but before that we should take a glance at the current state of the science.

The rise of resurrection science

First attempts and back-breeding

In antiquity when the remnants of historical species were exposed and the species were conceived of as non-existent, it was natural to pose a counterfactual question: if the species X still existed, what would the world be like? Because the new possibility of de-extinction could now be just around the corner – or so some optimistic proponents claim – we face a somehow similar prospective question: if the species X comes to exist again, what will the world be like then? In the public awareness and in popular culture, this question mostly concerns charismatic megafauna. The more academic debate on the question focusses on (in terms of geological epochs) more recently disappeared species that are also seen to have conservation value.

Probably the first science-based resurrection attempt was undertaken by German brothers Lutz and Heinz Heck in the 1920s and 1930s. They were zoologists and keen on reviving the aurochs that became extinct in 1627 and the tarpan that became extinct in the late nineteenth century, the wild ancestors of the cow and the horse, respectively. Because the brothers were active Nazis with close contacts to the party elite, and because they portrayed their efforts in overtly ideological terms, the Heck cattle have often been dubbed the Nazi cows (van Vuure 2005, pp. 323–59; Levy 2011, pp. 205–9). Today their technique is known as back-breeding, a form of selective breeding in which the starting point is the existing variety of domestic animals. The main difference between back-breeding and other, more common, forms of selective breeding is that in back-breeding the intention of the breeder is to reverse the course of breeding processes and bring back the qualities lost in domestication, whereas usually the aim of selective breeding is to bring about novelties by enhancing desired qualities such as increased milk production. As is demonstrated by Bart Gremmen and Christian Gamborg (in Chapters 7 and 3), it is usually impossible to put the evolutionary clock backwards; the Heck cattle morphologically differ from their wild ancestors. Thus, even though commonly used, the term 'back-breeding' should be considered an oxymoron except in some cases such as the following quagga project.

The Heck brothers' endeavour inspired South-African taxidermist Reinhold Rau (1932–2006) who took up reviving the quagga, a subspecies of zebra that became extinct in the late 1870s. Rau was able to show that, against the prevailing belief, quagga and the extant plains zebra are so closely related that back-breeding is a feasible option for the recreation of the former. The back-breeding of the zebra into quagga started in 1987. The focus has been on its visible characteristics, colour and the extent of striping, and by 2005 a fourth-generation offspring called Henry is dissimilar from plains zebra by appearance, in particular the loss of stripes at its hind quarters (the Quagga Project 2013).[2]

Genomic technology

Roughly, there are three main ways to resurrect extinct species: back-breeding, cross-species cloning and genetic engineering (Sherkow and Greely 2013, p. 32). The success of the last two rely, first, on the availability of sufficiently well-preserved tissue or some other source of genetic information, and, second, on the development of cloning, DNA synthesizing, genome reconstruction (Zimmer 2013, p. 37), stem cell technology (ibid., p. 36) and other related technologies. Luckily for de-extinction projects, preserving animal tissues in different ways (in cryobanks and 'frozen zoos', for example) is customary in biology. The practice has often been justified with a reference to biodiversity conservation, though researchers have saved biomaterial even without knowing whether it will be of any use in the future (Pask et al., 2008). This material is now thought to form a basis for the de-extinction of animals from which samples have been saved.

When biomaterial has not been preserved, as often is the case with long extinct animals, genetic material must be found and extracted from 'nature's own biobanks' – such as permafrost carcasses of the mammoth in the arctic area. DNA extracted from this kind of source is called ancient DNA, or aDNA for short, and its extraction is a disputable issue in science, since it can be damaged or contaminated.

The attitudes to the possibility of resurrecting extinct species from genetic material by cross-species cloning or genetic engineering have varied from doubt to wishful thinking and, following the story of scientific and technological progress, to expectation:

> We are a long, long way from being able to reconstruct the DNA of extinct creatures, and in fact it may be impossible to resurrect the DNA of dinosaurs or other long-extinct forms. We have DNA for living creatures, including ourselves, and yet we cannot clone any living

> animal (from DNA alone). As for extinct taxa, it is unclear whether or not DNA actually can survive more than a few thousand years. No one has been able to demonstrate incontrovertibly, as of yet, that they can retrieve DNA from an extinct species. (Horner 1999)

Richard Stone, the news writer for the prestigious journal *Science*, concludes his book on mammoths less doubtfully: 'whether it is five years, five decades, or five centuries from now, woolly mammoths will once again walk the earth' (Stone 2003, p. 215). A few years later in *Nature*, an article entitled 'Let's Make a Mammoth' (Nicholls 2008) concluded that over the next five decades de-extinction could be possible. In 2013, a cover story about de-extinction in *National Geographic* considers it to be an implementable technology: 'the revival of an extinct species is no longer a fantasy' (Zimmer 2013, p. 28). At the same time, the front page of the website for the research programme Revive & Restore proclaims:

> Genomic technology and techniques are advancing rapidly. It is becoming feasible to reconstitute the genomes of vanished species in living form, using genetic material from preserved specimens and archeological artifacts. Some extinct species may be revivable. (Revive & Restore 2013)

The articles in this volume are by and large less optimistic about the success of de-extinction, the reasons for which are not technological but essentially philosophical.

Let us take a closer look at the woolly mammoth. Most populations of the woolly mammoth disappeared 11,000 years ago, though there were some isolated populations that lived on, the last of which remained on the Artic island of Wrangel until they became extinct less than 4,000 years ago. The first ideas of resurrecting the species go back to the 1960s (Mayor 2011, p. 247). More recently, after the collapse of the Soviet Union in 1991, a few international expeditions have sought and found the remains of the mammoth so as to obtain its DNA. By 2012, the researchers were able to extract DNA from the tooth, bone and soft tissue of the mammoth (Paijmans et al. 2013), but neither a complete genome nor the DNA has been obtained intact (Loi et al. 2011, p. 229). A great deal of hope was put on some liquid blood that contained part of a carcass that was found in June 2013 (Switek 2013). However, for cloning purposes the DNA should ideally be uncontaminated and undamaged, but finding such may be virtually impossible (Nicholls 2008, pp. 310–11) and even the liquid blood found is unlikely to meet the needs (Switek

2013). Because of this, genetic engineering steps in and scientists look for possibilities to repair and complete the degraded genomes (Redford et al. 2013). Suggestions for reconstructing the genes have also been made. In any case, the DNA constructed would have to be inserted into elephant eggs or stem cells. This would require harvesting eggs from an elephant and an elephant 'surrogate mother', both of which are great challenges (Nichols 2008; Zimmer 2013, pp. 36–7).

The woolly mammoth is not the only possible species to be re-created by gene-technological applications. Abundant candidate species include: the thylacine (the Tasmanian tiger) (Fletcher 2010), the passenger-pigeon and the Carolina parakeet (Revive & Restore 2013), the moa (Huynen et al. 2012), the Yangtze River dolphin (baiji) and the aurochs (van Vuure 2005). There has been speculation over the reviving of *Homo neandertalis*, which, of course, raises even more serious moral concerns (Zorich 2010; Wynn and Coolidge 2012, pp. 181–2). The actual list of potential targets is far longer. Some of the de-extinction work is being done behind closed doors for commercial reasons: for instance, film producers have invested in certain projects with a wish to be the exclusive witnesses of the rebirth of the woolly mammoth (Stone 2003).

According to the *Scientific American*, the first successful conservation cloning was conducted in November 1999 when a gaur calf was born (Lanza et al. 2000). The gaur was not, however, extinct but an endangered species. The first extinct species to be revived was the Pyrenean Ibex (bucardo) in 2003, but the success was far from complete: numerous attempts resulted in only one birth of a morphologically normal female that died soon after birth (Folch et al. 2009). In Australia, the two existing species of gastric brooding frogs vanished in the 1980s, and since 2008 palaeontologist Michael Archer has tried to revive one of them and has succeeded in producing blastocysts that survived a few days. These species, which were only found in the early 1970s, have peculiar qualities because they were able to transform their stomach into a womb (Stone 2013, p. 1371). In Iran, a group has been able to create kinds of Esfahan mouflon lambs by coalescing cryobanked fibroblasts of a mouflon and a domestic sheep. The lambs, however, died soon after birth (Hajian et al. 2011). A Japanese group, led by Sayaka Wakayama, could hold the key to the viability problem. In 2008, they produced a healthy mouse from a frozen body by cloning, and in March 2013 the same group reported having recloned viable mice over 25 generations (Wakayama et al. 2013). It is, nevertheless, one thing to clone a laboratory mouse, another thing to revive a recently extinct animal and yet another thing to revive an authentic ancient animal from its ancient DNA.

Even though approaches to de-extinction have been divided into ones based on back-breeding and others carried out by genomic technologies, in practice the different methods may be mixed and used to complement the same project at different phases. For example, one of the possibilities to revive the woolly mammoth is a mixture of cross-species cloning and back-breeding. Cross-species cloning forms the first step, in which the nuclear DNA of the produced animal is from mammoth, mitochondrial DNA and other cellular machinery from an Asian elephant working as an egg donor and a surrogate mother (see Derek Turner and Christian Gamborg in Chapters 2 and 3). In the second step of the project the animals born from cloning are used for back-breeding in a way that aims to enhance the mammoth-like features and diminish the elephant-like features in order to gain animals that after each generation are more and more like mammoths.

Genetic modification of endangered species and mixing technologies

While numerous research projects reaching towards de-extinction are active and ongoing, at least to our knowledge, no research group is working towards the goal of genetically modifying endangered animal species to be more likely to survive and reproduce in a new or changing environment. Thus, at least this far, the genetic modification of endangered animals is merely an interesting idea, not an actual topic of biological research. Nevertheless, the idea is not as far-fetched as it first seems. When it comes to plants, genetic modification is constantly suggested and commonly studied as a method to make them more able to tolerate changes in their environment – especially changes such as extreme temperatures, drought, salinity and chemical toxicity following from climate change and other anthropogenic impacts (see for instance Wang et al. 2003; Eepen and D'Souza 2005; Velliyodàn and Nguyen 2006). This kind of modification is principally done to crop plants but also to forest trees, which are seen as 'central components of the global ecosystem' and thus central targets of conservation (Wang et al. 2003, p. 1). The idea to modify endangered animals could stem from this kind of modification of plants.

The idea of genetic modification aiming to increase changes for survival can be and has been connected to the idea of de-extinction. Some animals, Yangtze River dolphins for example, under de-extinction projects have died out from pollutants and other anthropogenic changes in their habitats (Zimmer 2013, p. 41). In other cases, such as some Hawaiian bird species, the extinction is caused by a pathogen still present in the habitat (Revive & Restore 2013). If these dolphins and birds are re-created and reintroduced into their historic ranges, they are likely to die out soon, and the whole exercise of de-extinction becomes futile. Thus, it has been suggested (Rosen 2012; Revive & Restore 2013) that re-created animals were genetically modified to tolerate their environment better. Another possibility is to control pathogens or pests that caused the extinction in gene-technological ways (for example by genetically modified viruses). If these suggestions are taken seriously, the justification for discussing questions concerning de-extinction together with questions regarding the genetic modification of endangered species and other forms of gene-technological conservation is obvious and strong.

As the above examples of Yangtze River dolphins and Hawaiian birds indicate, the stories of the resurrection science and the gene-technological conservation of endangered species are badly defective if limited to the work in laboratories. Species re-creation and gene-technological conservation are closely connected with the intentional alteration of ecosystems by means of species introduction. Thus, in order to be a realistic option, every de-extinction project has to include a systematic study of the biophysical condition of sites where the animals can live and reproduce. In other words, de-extinction requires resurrection ecology. For some biologists, the resurrected species, in particular vertebrates, could adopt a key role in an ecosystem and thus contribute to 'rewilding' it. One of the places already used for 'resurrected species' is Oostvaardersplassen in the Netherlands. This is the brainchild of Dutch ecologist Frans Vera. Oostvaardersplassen is a polder where Konik horses and Heck cattle dwell (see Bart Gremmen in Chapter 7). Visionary scientists Paul S. Martin (1928–2010) in the USA and Sergey Zimov (born 1955) in Russia have put forward ideas of restoring Pleistocene ecosystems in the Great Plains and in the Siberian Tundra, respectively (Stone 2003; Martin 2005; Zimov 2005; Donlan et al. 2006; Svenning 2007; Levy 2011, pp. 192–6). Martin's proposal has not been tested yet, apparently because of its controversial nature (see Rubenstein et al. 2006): it involves translocation of extant African animals to the prairie. Zimov has conducted experiments at the North-East Scientific Station in Yakutia since 1988. According to the project website, the aim is to restore 'the mammoth steppe ecosystem…The initiative requires replacement of the current unproductive northern ecosystems by highly productive pastures which have both a high animal density and a high rate of biocycling'(Pleistocene Park 2013; see also Zimov et al. 2012).

Zimov's grand-scale plans are probably the most meticulous there is for extinct species. Were the mammoth resurrected, it would have a place to roam, at least – which is rather a comforting prospect. The situation could be similar with the quagga. As noted above, regarding some other species, the situation can be less comforting; for example, the Australian gastric brooding frogs died out recently and so one may wonder whether the conditions for their second arrival are any better than they were at the time of their extinction.

For (some) evolutionary geneticists, palaeontologists and conservation biologists the goal of the revivification is as alluring as the Higgs boson has been for cosmologists. What is even more important to the feasibility of the goal is that this idea attracts many wealthy lay persons, the media and the entertainment industry (see Stone 2003) to such an extent that more and more effort is put into studying extinct species and their remains. In short, we are witnessing the rise and the formation of a new field of scientific inquiry known as 'resurrection science' (Judson 2008; see also Derek Turner in Chapter 2). With the expression 'resurrection science' we refer to the whole range of efforts to provide systematically a new life to extinct species (or provide changes to enable survival for endangered species) by means of scientific inquiry and related technologies. This term is deliberately ambiguous and can be used in narrower and broader ways. In the narrowest sense, resurrection science covers the inquiry and the development of techniques that are driven by the explicit aim to bring some extinct species back to life. In the broadest sense, the expression denotes also the gene-technological conservation of endangered species as well as basic palaeontological inquiry and the development of genomic technologies, such as genetic sequencing and genetic engineering. These scientific enterprises provide understanding about, and the means for, implementing de-extinction and the genetic modification of endangered animals without it being the specific goal.

Problematizing de-extinction and gene-technological conservation of endangered species

Even though there is forceful research on, and optimism about, resurrection science, many issues remain open and deeply controversial, not least from the philosophical perspective. Most fundamentally, the idea of de-extinction of species is not conceptually as clear, nor is its moral desirability as self-evident, as its advocates suggest. Similar questions rise with respect to the genetic modification of endangered animals as well as other forms of gene-technological conservation. In this section, we will shed light on crucial philosophical and ethical questions regarding resurrection science and genetic modification. To begin with, we will consider the meaning and possibility of the anthropogenic revival of species. Strictly considered, if it turns out that the metaphysical foundations of resurrection science are flawed, it entails an end to further reflection about it and that it is close to being a pseudo-science. For the sake of argument, we will assume, however, that talk about resurrection is meaningful. Then we can face the questions about the ethical grounds and the nature of resurrection science.

Is reviving possible?

Is it, even in theory, possible for resurrection science to carry out its promise and bring some extinct species back to life? Even though the field is developing fast and the results look outstanding, there are philosophical issues that pre-empt satisfaction: the animal born from the de-extinction procedures may not be the kind of animal that its creators claim it to be. This is a question about the species identity of animals brought into existence (Ehrenfeld 2006, pp. 730–1; Zimmer 2013, p. 41; Redford et al. 2013, p. 3). What we have called resurrection science in the narrow sense is founded on highly specific assumptions about the nature of species. These assumptions form a species concept that is genome-centred, morphology-centred or both. As the proponents of de-extinction see it, a species consists in genomes and/or a species membership that is determined by a morphological similarity with other (past) members of the species. An animal's lack of ecological interactions with its (native) biophysical environment, social relations and the way it came into existence is seen as insignificant to its species identity.

Genetic reductionism (sometimes also genetic determinism) is an idea that the development and identity of an organism is solely determined by its genes. It means that if we have the whole genome from an extinct animal and sufficient technology to re-create a living animal that is genetically identical to that extinct animal and capable of reproducing itself, then we have succeeded in recreating the species. However, a phenotype of an organism does not solely result from its genes. The development of an organism depends on its environment and the molecular mechanisms it triggers at the genomic level (see Julien Delord in Chapter 1; Blumberg 2010, pp. 44–5; Carey 2012). When an animal develops in the womb of a surrogate mother of another species and lives its whole life without any contact with members of its own genetic species or outside the natural environment of its genetic species, it is justified to ask whether that animal (despite its genetic similarity to members of a certain species) fails to be a member of its genetic species (see Julien Delord and Helena Siipi in Chapters 1 and 4). It is possible that the answer varies from one species to another. Some animals live in groups and others virtually alone. Moreover, the behaviour of some animals is mostly instinct-based, whereas the species-typical behaviour of other animals results from learning. Thus, it may be more representative to classify a solitary animal in the species of 'instinct-driven hermit' than as a lonesome member of a more social animal species.

The view that being a member of a species consists of genetic, morphological and behavioural similarity to other members is far from self-evident, as there is neither a universally accepted criterion of species membership nor a universally accepted concept of 'species' (de Queiroz 1998, p. 57; Powell 2011, p. 604). Even though some concepts of 'species' cohere with the idea of de-extinction, others may not, which becomes clear from the following example of philosopher Brian Garvey:

> On the cladistics conception of taxonomy, which is the currently orthodox view, it is a necessary condition for something being a member of the species *Equus caballus* that it shares a common ancestor with all and only members of that species. Thus, no matter what properties a creature possesses – no matter how much it looks like a horse, walks like a horse, neighs like a horse – if it does not have the ancestry common to horses, it is not a horse... Thus, if the horse species goes extinct, but creatures that are exactly the same as them come into existence later on, those creatures will not be horses. (Garvey 2007, p. 150)

What Garvey wants to highlight is that belonging to the same species presupposes a connection or 'a causal contact' between animals. Animals born from back-breeding, cross-species cloning or genetic engineering, of course, have some genetic connection to the species they are claimed to belong to and with which they are more or less genetically similar. It is not clear, however, whether that connection is strong enough to justify species membership. If the answer is negative, then the animal produced is an individual that falls outside the conventional species taxonomy and thus there is no real or genuine de-extinction. The plausibility and the acceptability of the idea of de-extinction depends on spelling out a justified positive answer to the question. Alternatively, one would have to argue for the acceptability of the genome and/or morphology-centred conceptions of 'species'. The answers given to these questions concerning species identity vary according to the de-extinction method and the sources of genetic material. In any case, the views regarding the identity of animals born from de-extinction are the starting point of any further queries concerning de-extinction. Questions concerning conceptions of species and species identity are discussed in several chapters in this volume (especially in 1 and 4). As long as the status of species of genetically modified animals is under philosophical discussion (see for example Lee 2003; Savulescu 2011), similar questions arise and these are equally relevant to the idea of preserving endangered species by genetic modification.

The possibility of de-extinction further depends on the conceptions and views about extinction. The concept of extinction applies to species, but it is often seen to be analogous with the concept of death that applies to individual organisms (Gunn 1991, p. 299; Delord 2007, p. 657). Words often used to describe de-extinction – reviving, resurrecting, waking from death – bridge the two ideas. The terms 'death' and 'extinction' are both sometimes seen to carry the idea of irreversibility with them, but in both cases this view has been contested (Cole 1992; Delord 2007). Thus, discussion about the possibility of de-extinction is linked with a wider philosophical discussion on the irreversibility of death. The possibility of re-creating extinct animals implies abandoning the received wisdom in conservation biology that the loss of a species is its end. Alastair Gunn (1991, p. 299) has asked whether this wisdom is merely based on our limited technological abilities or whether the finality should be seen as something conceptually essential for extinction. If resurrected animals after all turn out to be real members of the original species, should we say that the resurrected species was only temporarily extinct? Or should we hold on to the idea of the finality of extinction, and argue that as long as there is at least a distant possibility of re-creation, the species should not be considered as extinct (maybe it should be seen as temporarily extinct or potentially existing)?

If possible, is it acceptable?

Supposing that the animals born from de-extinction and genetic modification are real and authentic members of the original species. In this case we would have to ponder over the reasons for and against the adoption of these technologies. Furthermore, we would encounter questions about the moral significance of the possibility of resurrecting extinct animals. The answers given at least partly depend on what is done with the animals once they are brought into existence. A lonesome individual or a small group of animals living in a tight human control raises concerns that differ from those that the populations of re-created and/or genetically modified animals in the wild raise. Although the conservational value of the latter possibility is potentially higher, it also, as argued by Myhr and Myskja in Chapter 6, carries risks and questions foreign to the former alternative. Animals living in zoos and laboratories, on the other hand, raise welfare questions foreign to those living in the wild. Nevertheless, some ethical questions concern both alternatives.

Fundamentally, the question about acceptability comes down to two questions. First, what are the reasons for de-extinction and the genetic modification of endangered species? Second, what could be wrong in bringing animals into existence by de-extinction or in preserving their existence by different forms of genetic modification?

Reparation and biodiversity

Most people consider extinction of a species as a prima facie bad thing or as a disvalue. Consequently, it is intuitively sound to claim that a state of affairs in which a species N exists is prima facie better than a state of affairs in which the same species does not exist. From these statements it seems to follow that, if extinction can be reversed, there are intuitive moral reasons for reversing it. De-extinction and the genetic modification of endangered species can be thought of as a form of reparation and thus involving the positive human duties in regard to biodiversity (see Chapters 2, 3 and 7). Through these actions we could reduce the harm and redress the destruction caused by past human behaviour. As noted on the website of the Quagga Project, 'the project is aimed at rectifying a tragic mistake made over a hundred years ago through greed and short sightedness' (Quagga Project 2013).

The goal and idea of reparation is deceptively easy to connect with the Pleistocene extinctions. The mammoth, the sabre-toothed tiger, the Irish elk and other big mammals disappeared rather quickly when the ice cover retreated northwards. One of the conjectures and explanations offered for these extinctions is the so-called overkill hypothesis that was first developed by Martin in the 1960s (Levy 2011, pp. 4, 19–26) and to which Zimov et al. (2012) subscribe. According to the hypothesis, Stone Age hunters eradicated the Pleistocene megafauna and destroyed the Pleistocene ecosystem within the period of a few generations. The process of human-caused destruction has continued ever since:

> Climatic change is always of interest but not crucial in formulating explanations. As our species spread to various continents, we wiped out their large mammals; as we progressed to oceanic islands, we extinguished many mammals that were much smaller, and even more birds, especially flightless species. (Martin 2005, p. 48)

Although the drivers of most famous historical extinctions (for instance the loss of the Pleistocene megafauna) still remain a subject of scientific dispute, nobody today denies that the number of human-induced losses of species is on a steep rise. Therefore, the idea of reparation is certainly ethically appealing. At the same time, it carries with itself a radical overtone: the idea that conservation should consist of active species introduction and modification of the living world around us. Thus it paves the way for a new understanding of human-caused damage in nature. In contrast to the mere 'hands off' approach, these terms refer to active human participation and intervention in the natural processes. Most importantly, the notions of repairing and compensating denote a category of actions for which there are morally strong reasons due to wrong action in the recent past. However, since it is unclear how the term 'damage' should be understood in the environmental context, the ideas of reparation, compensation and harm reduction also require further analysis. Most of all, we have to consider how they differ from other anthropogenic changes and what the relationship between the resurrected animals and the extinct forerunners is like.

The idea of reparation is not as easily acceptable as it first sounds, or at least it is restricted to certain species. First, the flipside of evolutionary speciation is that unviable varieties are wiped out and that, in geological terms, most species that have ever lived have become extinct. Second, there are species that we wish to see disappear because they are harmful to us. Third, it can be the case that only those extinctions (and not even all of them) in which humans are involved are morally suspect. If so, the sixth mass extinction wave is the main worry here, and pre-anthropocene extinctions and de-extinction projects like *Jurassic Park* fall under a different moral category. Fourth, the idea of reparation rests on the idea that species as historical collective entities have value. Does the value of resurrected species equal the value of (natural) species? Are genetically modified 'wild' species as valuable as the ones that are freer from human-made modifications? These questions will be discussed by Elisa Aaltola and Markku Oksanen in Chapters 5 and 8.

Anthropocentric reasons for de-extinction and genetic modification

Reasons for de-extinction and genetic modification can be far from altruistic. The motivation concerning endangered species can stem from the valuable services and resources these animals can provide for human beings. At the minimum, re-created animals are likely to satisfy scientific curiosity and provide 'a unique opportunity to study living members of previously extinct species (or at least close approximations of those species)' (Sherkow and Greely 2013, p. 33).

Some re-created and endangered animals could assist in the reaching of new scientific discoveries and useful products. It has been suggested, for example, that bringing back the gastric brooding frog that can turn its stomach into a womb could help in developing new infertility treatments (Zimmer 2013, p. 40). For some people, the idea of seeing a live mammoth is simply attractive; thus, re-created animals could provide us with fascinating experiences. But how good are these reasons? In Chapter 2, Derek Turner discusses them and offers a further argument based on the idea of ecosystem health.

Animal well-being

Sometimes the goals of biodiversity conservation and the goals of promoting the good of individual animals contradict each other. The killing of individuals is often used in protecting extant biodiversity, as in the case of the control of invasive alien species. De-extinction and the genetic modification of endangered species is also a hot topic in this respect. Animals brought into existence by de-extinction and genetic modification, as well as animals used for producing them (surrogate mothers and egg donors for example), might suffer from the methods and processes used. Particular genomic variations may bring about deformity and untimely death (Sherkow and Greely 2013, p. 32). As mentioned above, so far endeavours at revival have been rather costly when considered in terms of untimely deaths and welfare. In most countries, animal welfare legislation constrains scientific enterprises in this area and forbids unnecessary suffering. The best known theories of animal ethics – those of Peter Singer and Tom Regan – are thoroughly individualistic. For both of them, it is morally wrong to base the differential treatment of sentient animals solely on their species membership.

Resurrection attempts can cause unnecessary suffering of animals because of poor living conditions. As many animals would have to live in zoos and laboratories, they may suffer from the lack of company, the tedious environment and extensive human curiosity (ibid., p. 19). Lack of conspecifics or normal social environment could also cause suffering. Most writers in this book pay attention to the possible harm and suffering that resurrection science is likely to cause.

Ecological risks

The acceptability of de-extinction and genetic modification partly depends on the outcome of risk–benefit analysis. The question is: what risks are involved with these methods, beside welfare concerns?

According to David Ehrenfeld (2006, p. 724), biotechnological means in conservation are making the identification of risks difficult. He assumes that many risks are similar to known risks in agriculture. He continues by noting that the best known risks of genetically modified organisms concern transgenes escaping into the wild relatives of the species. This worry is especially topical with respect to de-extinction produced by genetic engineering and cross-species cloning (and possibly combined with genetic modification for better survival). These de-extinct animals, or at least the first generation of them, will live in close contact with the surrogate species and even share some of their genes with them. Should we then worry about gene flow from the de-extinct species to the surrogate one? Most likely the spread of transgenes is, of course, in the genetic modification of endangered species. The transgenic animals are then modified for better survival. However, in their case, the spread of transgenes may not be problem; rather it could be considered as a desired outcome.

A further worry is that a de-extinct species or a genetically modified species would not only be successful in its new environment, but could develop into an invasive species harming other species and ecosystems (ibid., pp. 728–9; Rosen 2012; Stone 2013, p. 19; Redford et al. 2013, p. 3). A further risk is that animals born from de-extinction could be a source of harmful diseases (Stone 2013, p. 19; Zimmer 2013, p. 41). Sherkow and Greely (2013, p. 19) state that 'newly de-extinct creatures might prove excellent vectors for pathogens. An extinct animal's genome could also conceivably harbour unrecognized endogenous retroviruses.'

The risks, and probably also the benefits, certainly differ from one species to another. Nevertheless, as Myhr and Myskja propose in Chapter 6, ways of managing and assessing risks must be an integral part of any de-extinction procedure as well as the genetic modification of endangered species.

Moral hazard

In the same issue of *Scientific American* in which the reviving of the gaur is reported, the editor reflects on the larger worries that are raised:

> Forgive my paranoia, but I can imagine a future time in which a land-use developer argues that there is no reason to worry about the disappearance of a given species in the wild because we can always resurrect it later through cryogenics and cloning – whereas we need that ranch land now. (Rennie 2000, p. 1)

The editor formulates here the problem of a moral hazard that is widely known and studied in insurance economics: having an insurance policy increases recklessness in behaviour. Applied to conservation, when re-creating animals becomes technically possible, the received wisdom in conservation biology loses its appeal. According to it, the loss of a species is irreversible and eternal. Thus, the development of new technology may significantly weaken the motivation for *in situ* conservation.

The hazard of recklessness is partly empirical, partly theoretical. It is a matter of empirical study to monitor whether the introduction of new conservation possibilities really increases losses of biodiversity. In the context of conservation cloning, it is obvious that the technology is the last resort for prolonging the life of an endangered species or is the only possibility for reviving it. The hazard of recklessness is discussed in numerous chapters in this volume, in particular in 2 and 4.

* * *

There are no simple answers to the issues of using advanced technologies in biodiversity conservation in the complex non-ideal world, nor are the authors of the following chapters offering them. Rather, in this book we wish to provide the reader with the conceptual means to assess critically the promise and the prospects of resurrection science.

Notes

1. Today, the search for surviving populations or individuals of 'officially' extinct and non-existent species is called cryptozoology. It is sometimes considered as a pseudo-science, since it involves searching for animals whose existence is highly unlikely, such as the Loch Ness monster. However, this is not a plausible view of the field. A great deal of biological fieldwork involves inventory work, such as the search for the last remaining individuals of the Yangtze River dolphin. Moreover, real animals tend to appear mythological as long as they are insufficiently documented and unknown to science. Through sufficient observations, this impression vanishes and the status of species changes from existence-not-proven to existence-proven. Examples of species that have gone through this kind of epistemological change include the Komodo dragon, the coelacanth and the platypus.
2. The project is still work-in-progress; both the history and the latest information on it, with the pictures of the new-borns, are available on the project website at www.quaggaproject.org.

References

Barrow, M.V. Jr (2009) *Nature's Ghosts: Confronting Extinction from the Age of Jefferson to the Age of Ecology* (Chicago: University of Chicago Press).

Blumberg, M.S. (2010) *Freaks of Nature and What They Tell Us About Development and Evolution* (Oxford: Oxford University Press).
Carey, N. (2012) *The Epigenetics Revolution* (London: Icon).
Cole, D.J. (1992) 'The Reversibility of Death', *Journal of Medical Ethics,* 18, 26–30.
Cuddington, K. and M. Ruse (2004) 'Biodiversity, Darwin, and the Fossil Record' in M. Oksanen and J. Pietarinen (eds) *Philosophy and Biodiversity* (Cambridge: Cambridge University Press).
Darwin, C. (1998) *The Origin of Species* (Oxford: Oxford University Press).
Delord, J. (2007) 'The Nature of Extinction', *Studies in History and Philosophy of Biological and Biomedical Sciences,* 38, 656–67.
Donlan, C.J. et al., (2006) 'Pleistocene Rewilding: An Optimistic Agenda for Twenty-First Century Conservation', *The American Naturalist,* 168, 660–81.
Eepen, S. and S.F. D'Souza (2005) 'Prospects of Genetic Engineering of Plants for Phytoremediation of Toxic Metals', *Biotechnology Advances,* 23(2): 97–114.
Ehrenfeld, D. (2006) 'Transgenics and Vertebrate Cloning as Tools for Species Conservation', *Conservation Biology,* 20(3), 723–32.
Fletcher, A. (2010) 'Genuine Fakes: Cloning Extinct Species as Science and Spectacle', *Politics and the Life Sciences,* 29, 48–60.
Folch, J. et al. (2009) 'First birth of an Animal from an Extinct Subspecies (*Capra pyrenaica pyrenaica*) by Cloning', *Theriogenology,* 71, 1026–34.
Garvey, B. (2007) *Philosophy of Biology* (Stocksfield: Acumen).
Gunn, A.S. (1991) 'The Restoration of Species and Natural Environments', *Environmental Ethics,* 13, 291–312.
Hajian, M. et al. (2011) ' "Conservation Cloning" of Vulnerable Esfahan Mouflon (*Ovis orientalis isphahanica*): In Vitro and In Vivo Studies', *European Journal of Wildlife Research,* 57, 959–69.
Horner, J. (1999) 'Scientific American. Ask the Experts FAQ File', www.scientificamerican.com/article.cfm?id=how-close-are-we-to-being (accessed 12 April 2013).
Horner, J. and J. Gorman (2010) *How to Build a Dinosaur* (New York: Plume).
Huynen, L et al. (2012) 'Resurrecting Ancient Animal Genomes: The Extinct Moa and More', *Bioessays,* 34, 661–9.
Jefferson, T. (1999 [1781]) *Notes on the State of Virginia,* ed. by F. Shuffleton (East Rutherford, NJ: Viking Penguin).
Judson, O. (2008) 'Resurrection Science', *The New York Times,* 25 Nov 2008, http://opinionator.blogs.nytimes.com/2008/11/25/resurrection-science/ (accessed 12 September 2013).
Lanza, R.P., et al. (2000) 'Cloning Noah's Ark', *Scientific American* 2000 (November), 849.
Lee, K. (2003) 'Patenting and Transgenic Organisms: A Philosophical Exploration', *Techné: Journal of the Society for Philosophy and Technology,* 6(3), 1–16.
Levy, S. (2011) *Once and Future Giants* (Oxford: Oxford University Press).
Loi, P. et al. (2011) 'Biological Time Machines: A Realistic Approach for Cloning an Extinct Mammal', *Endangered Species Research,* 14, 227–33.
Martin, P.S. (2005) *Twilight of the Mammoths: Ice Age Extinctions and the Rewilding of America* (Berkeley, CA: University of California Press).
Mayor, A. (2011) *The First Fossil Hunters: Dinosaurs, Mammoths, and Myth in Greek and Roman Times* (Princeton: Princeton University Press).

Mayr, E. (1988) *Toward a New Philosophy of Biology: Observations of an Evolutionist* (Cambridge, MA: Harvard University Press).
Meyer, A. (2005) *The DNA Detectives* (New York: Thunder Mouth Press).
Moore, J.A. (1999) *Science as a Way of Knowing: The Foundations of Modern Biology* (Cambridge, MA: Harvard University Press).
Nicholls, H. (2008) 'Let's Make a Mammoth', *Nature*, 456, 310–14.
Pääbo, S. (1985) 'Molecular Cloning of Ancient Egyptian Mummy DNA', *Nature*, 314, 644–5.
Paijmans, J.L.A. et al. (2013) 'Mitogenomic Analyses from Ancient DNA', *Molecular Phylogenetics and Evolution*, 69, 404–16.
Pask, A.J. et al. (2008) 'Resurrection of DNA Function *In Vivo* from an Extinct Genome', *PLoS ONE* 3(5).
Pimm, S. (2013) 'The Case against Species Revival', http://news.nationalgeographic.com/news/2013/03/130312–deextinction-conservation-animals-science-extinction-biodiversity-habitat-environment/ (accessed 5 April 2013).
Piña-Aguila, R.E. et al. (2009) 'Revival of Extinct Species Using Nuclear Transfer: Hope for Mammoth, True for the Pyrenean Ibex, But Is It Time for "Conservation Cloning"?' *Cloning and Stem Cells*, 11, 341–6.
Pleistocene Park (2013) 'Restoration of Mammoth Steppe Ecosystem', www.pleistocenepark.ru/en/ (accessed 26 July 2013).
Powell, R. (2011) 'On the Nature of Species and the Moral Significance of their Extinction' in T.L. Beauchamp and R.G. Frey (eds) *The Oxford Handbook of Animal Ethics* (Oxford: Oxford University Press).
Quagga Project (2013) www.quaggaproject.org/ (accessed 15 August 2013).
Queiroz, K. de (1998) 'The General Lineage Concept of Species, Species Criteria, and the Process of Speciation' in D.J. Howard and S.H. Berlocher (eds) *Endless Forms: Species and Speciation* (Oxford: Oxford University Press).
Redford, K.H. et al. (2013) 'Synthetic Biology and Conservation of Nature: Wicked Problems and Wicked Solutions', *PLoS Biology* 11(4).
Rennie, J. (2000) 'Editorial: Cloning and Conservation', *Scientific American*, 2000 (November), 1.
Revive & Restore (2013) http://longnow.org/revive/ (accessed 27 August 2013).
Rogers, C.P. (2002) 'Solution or Stumbling Block: Biological Engineering and the Modern Extinction Crisis', *Georgia Journal of International and Comparative Law*, 30, 141–63.
Rosen, R.J. (2012) 'Assuming We Develop the Capability, Should We Bring Back Extinct Species?' *The Atlantic*, 7 August.
Rubenstein, D.R. et al. (2006) 'Pleistocene Park: Does Re-wilding North America Represent Sound Conservation for the 21st Century?' *Biological Conservation*, 132, 232–8.
Rudwick, M.J.S. (2005) *Bursting the Limits of Time: The Reconstruction of Geohistory in the Age of Revolution* (Chicago: University of Chicago Press).
Savulescu, J. (2011) 'Genetically Modified Animals: Should There Be Limits to Engineering the Animal Kingdom?' in T.L. Beauchamp and R.G. Frey (eds) *The Oxford Handbook of Animal Ethics* (Oxford: Oxford University Press).
Sherkow, J.S. and H.T. Greely (2013) 'What If Extinction Is Not Forever?' *Science* 340, 32–3.

Stone, R. (2003) *Mammoth: The Resurrection of an Ice Age Giant* (London: Fourth Estate).

Stone, R. (2013) 'A Rescue Mission for Amphibians at the Brink of Extinction', *Science*, 339, 1371.

Svenning, J.-C. (2007) '"Pleistocene Re-wilding" Merits Serious Consideration also Outside North America', guest editorial, *IBS Newsletter*, 5, 3–10.

Switek, B. (2013) 'Can Purported Mammoth Blood Revive Extinct Species?', *National Geographic: Daily News*, http://news.nationalgeographic.com/news/ (accessed 26 June 2013).

Taylor, P.W. (1986) *Respect for Nature: A Theory of Environmental Ethics* (Princeton: Princeton University Press).

Velliyodan, B. and H.T. Nguyen (2006) 'Understanding Regulatory Networks and Engineering for Enhanced Drought Tolerance in Plants', *Current Opinion in Plant Biology*, 9, 1–7.

Vuure, C. van (2005) *Retracting the Aurochs: History, Morphology and Ecology of an Extinct Wild Ox* (Sofia: Pensoft).

Wakayama, S. et al. (2013) 'Successful Serial Recloning in the Mouse over Multiple Generations', *Cell Stem Cell*, 12, 293–7.

Wang, W.B. et al. (2003) 'Plant Responses to Drought, Salinity and Extreme Temperatures: Towards Genetic Engineering for Stress Tolerance', *Planta*, 218, 1–14.

Wynn, T. and F.L. Coolidge (2012) *How to Think Like a Neanderthal* (Oxford: Oxford University Press).

Zimmer, C. (2013) 'Bringing Them Back to Life', *National Geographic* (April), 38–41.

Zimov, S.A. (2005) 'Pleistocene Park: Return of the Mammoth's Ecosystem', *Science*, 308, 796–8.

Zimov, S.A. et al. (2012) 'Mammoth Steppe: A High-productivity Phenomenon', *Quaternary Science Review*, 57, 26–45.

Zorich, Z. (2010) 'Should We Clone Neanderthals? The Scientific, Legal, and Ethical Obstacles', *Archeology* 63 (March/April), www.archaeology.org/1003/etc/neanderthals.html (accessed 27 August 2013).

1

Can We Really Re-create an Extinct Species by Cloning? A Metaphysical Analysis

Julien Delord

Introduction

During the last two decades, the idea that lost species could be re-created or resurrected from their fossil DNA with the help of adequate biotechnological machinery has become immensely popular. Even if all sorts of *Jurassic Park* fantasies, such as re-creating extinct dinosaurs, have been completely dismissed by scientific studies on the rate of DNA destruction, more realistic projects of fossil DNA sequencing are commonly conducted today. Indeed, depending on the conditions of conservation, it is generally accepted that, beyond a few tens of thousands of years, DNA strands are too degraded to be read – apart from the genomes of the many species which became extinct during the Holocene epoch (*c.*10,000 BC–present), more likely than not, as a result of anthropic pressures. Today the mammoth genome is almost completely deciphered (Miller et al. 2008) and other teams have chosen to sequence cave bear DNA (Noonan et al. 2005) or the Tasmanian tiger genome (Miller et al. 2009). And many more species will be added to the list in the coming years with the booming of 'museomics'. As a contraction of 'museum' and 'genomics', museomics refers to the new area of research by large-scale analysis of the DNA sequences preserved in specimens from museum collections all around the world.

The scientific relevance of sequencing fossil DNA today remains purely of cognitive interest without direct practical applications, except in helping taxonomists, palaeontologists and anthropologists to make sense of past biological events. Even though many palaeogeneticists assert that they do not intend to resurrect species and fix limits to their knowledge and technical capabilities (Orlando 2005), other powerful interests may be driven by scientific, ecological or nationalistic ambitions (Fletcher 2008) and may lobby actively for the implementation of such projects. Therefore, it seems wise and timely to look into the future, to explore the epistemological, metaphysical and ethical implications of a possible technology that could change our conception of life and evolution, even if such were to remain unrealized forever.

The scope of this chapter does not include the ethical dimension; instead, it focuses on the metaphysical aspects of species re-creation that I conceive as a prerequisite for axiological enquiries. More specifically, I will show that the question 'can we re-create an extinct species?' amounts to resolving what I call the 'resurrection paradox' and asking 'do organisms cloned from the DNA of an extinct species belong to this species?'. The answers will depend on the different metaphysical commitments about what species really are. I will begin by presenting the possible techniques for species re-creation and their scientific context.

How species go extinct and how they could be revived

A classification of species extinction

One cannot consider the possible re-creation of species without understanding firstly the different kinds of species extinctions. Usually, four different modalities of extinction are distinguished in an evolutionary context. In the first and the most obvious case, a species can purely and simply stop existing when no organism survives and reproduces: this is called 'phyletic' or 'final' extinction. A species can also vanish by hybridization with another interfertile species, and thus produce a new hybridogenous daughter species from the two mother species. This is for instance what happened to a fish species (the Whitefish) in Lake Geneva. The third way for a species to go extinct is that it transforms itself into a new species as a result of ecological or genetic modifications that produce an apomorphic trait; this kind of extinction is sometimes called – although inappropriately – 'pseudoextinction' (Van Valen 1973). Finally, a species can disappear by giving birth to two or more daughter species in a process of allopatric speciation.

It is worth noting from the outset that people get excited by the possible re-creation of totally extinct species, and not species that were only submitted to a process of anaphyletic change or speciation. The reason is that the return to life and the development of an additional phylum in the tree of life, the survival of an original process of evolution, and the transmission of certain genetic features and information are really valued from an ecological point of view.

However, although less spectacular and appealing to conservationists, the cloning of old specimens from a species still alive could be a very important, even a necessary preliminary, experiment before trying to achieve an authentic resurrection. It could be easier to achieve with greater chances of success and it could serve as a basis of comparison for complex interspecies cloning experiments. Ancient DNA sequences are of huge theoretical importance for inferring temporal changes in phyla and assessing the rate of genome evolution, as stated above. But the genome is not the mirror of the organism. Only a living organism can provide some of the precious pieces of non-reducible biological information such as certain behavioural or ecological interactions. These then become available for direct comparison with the species in the present.

Here, the difference between re-creating a species and resurrecting a particular organism becomes evident. Nobody really hopes – except during psychoanalysis – to re-create an eight-year-old murder victim, though everybody wishes that the child could be resurrected.

A biotechnological recipe for resurrecting species

The basic idea of biotechnological species resurrection goes like this. Re-create the full genome from one or several organisms that belong to a now extinct species; encapsulate it in a nucleus (natural or synthetic); inject it in an enucleated receiver ovum from a phylogenetically close species; let the new organism develop in the reproductive organs of a surrogate female from this close species. This kind of biotechnological manipulation was achieved in 2010 by a Craig Venter Institute's team which created an artificial bacterial genome that was successfully introduced in a host bacterium whose own chromosome had been previously removed. But it is a long way from the creation of this artificial bacterium (called 'Synthia') to the cloning of a eukaryotic cell from an extinct species, such as a mammoth or a moa (Gibson et al. 2010).

Technically speaking, this clone is not a perfect reproduction of the dead organism belonging to the extinct species. Actually, it is a nucleo-cytoplasmic chimera, and the cellular machinery as well as the mitochondrial genome remain those of the bearing species. Hopefully, the descent of these clones could be improved to incorporate more features of the extinct species, beginning with its mitochondrial genome.

More generally, a recent paradigm change in the modern evolutionary synthesis toward the integration of what is called 'soft inheritance', as opposed to genetic heredity, forces us to consider the role of epigenetics in all its dimensions (chromatin structure, cellular composition and so on), including cultural evolution as well as niche inheritance, in the definition of what is transmitted by species membership (Danchin et al. 2011). If, in the long run, the main support of heredity is still considered to be DNA and the genes that are coded on it, in the short term, from one generation to the next, many different kinds of cultural, ecological as well as epigenetic *variants* are transmitted, so much so that evolution has been recently redefined as 'the process by which the frequencies of variants in a population change over time' (Bentley et al. 2004, p. 1443). Cloning an extinct specimen out of its genetic features only, with allogenetic specifications concerning its 'soft' heredity, is clearly misleading with regard to the objective biological parameters of the populations and the organisms composing the species that became extinct. However, if their range of variation is maintained under certain limits, so that they do not trigger the development of new characters that could be considered as apomorphies, the cloned organism should presumably be considered as belonging to the same species as its genetic template.

I will not discuss in detail all the empirical obstacles standing in the way of the achievement of an effective cloning of an extinct species; the point is to underline the theoretical possibility of the re-creation of an organism belonging to an extinct species provided that the following technical requirements are satisfied:

- It is necessary to have a significant part of the genome of at least one individual from species S (or two, male and female, for two-sex species) and the biotechnologies allowing for the expression of this genome in an appropriate cellular machinery that is comparable to that of the extinct species S.
- Genetic information should be retrieved, sequenced, stored and transmitted without alteration.
- The artificial (biotechnological) and the natural expression of genetic information are equivalent.

The resurrection paradox and the metaphysics of species

The list of technical requirements is independent of the philosophical requirements for the re-created clone to belong to an extinct species. These are metaphysical arguments about the nature of species, of change, and of human technological action. The meaning of biotechnological modifications intrinsically depends on how we interpret events such as extinction, cloning, evolution and the relationship between genetic information and the identity of a species.

The question of species resurrection lies at the crossroads of many conceptual difficulties in biology, notably about the species problem. Thus, we start from a given *product* of the evolutionary history (an extinct species or an individual organism belonging to an extinct species) and we want to end up with a renewed evolutionary *process*, making the previous extinct species again a functioning, cohesive and evolving species. This being more than a technical difficulty, I call this the 'resurrection paradox'.

Indeed, if one thinks in terms of products, the so-called extinction is just a temporal gap between the occurrences of two organisms (products) of the species, and which is not a real definitive extinction. This amounts to saying that there is no real resurrection either. Alternatively, if one goes with the process perspective, the species went definitively extinct at the end of the reproductive or living process, and so there is no resurrection at all: we only witness the results of a human modification on the living process of the bearing species, which is another stream of life than the extinct one.

In order to find a way beyond this resurrection paradox, we have first to go back to the origins of the two conflicting positions, the never-ending dispute between two unresolvable metaphysical positions in the philosophy of biology: the *species-as-class* (or natural kind) and the *species-as-individual* stances. Reydon (2008, p. 166) reminds us that this debate amounts to an ontological dispute regarding the relation between organisms and the species to which they belong: 'on the class view this is a relation of membership, while on the individuals view organisms are parts of their species in the same way as my cells are parts of me'.

We should also subdivide each position into distinct currents. Following the literature on the subject (Okasha 2002; Stamos 2003; Reydon 2008), it is possible to distinguish at least two types of class depending on the nature of the property defining the characteristics of membership: the members of a given class possess a real essence; and the members of the class exhibit a relational essence with no necessary property, only some sufficient ones. On the other hand, it is now current to distinguish two concepts of individuality relative to change and time metaphysics: vertical and horizontal in biological terms or presentist/eternalist in metaphysical terms. In the first case, the existing material objects only exist in a three-dimensional space with time flowing independently; in the second case, objects fully exist in a four-dimensional space–time universe and are always 'present' in their past, present and future dimensions.

These slight nuances happen to be quite important when dealing with the question of extinction and biotech ecology because they introduce diverging positions when confronted with the identity of species through time as well as about the nature of processes, be they evolutionary, ecological or genetic, in the characterization of species.

Real essentialism

I will begin my review with the two simplest positions: the one that fully agrees with the possibility of species resurrection and the one that makes it conceptually impossible.

The one that fully permits 'awakening the dead', at least at the species level, is real essentialism, directly derived from Aristotelian metaphysics. This kind of classical essentialism states that all the members of a species are fully members of this species and only of this species because they all share a property or a set of properties, which is their essence. A horse is essentially a horse because it shares with all the other horses the property of 'horseity', something that Aristotle and his contemporaries equated with *specie differentia*, a difference affecting the four causes (material, efficient, formal, final) explaining the existence of horses. Although essentialism has had to face many rebuttals in biology, especially since Darwin and the demonstration that species are not fixed and immortal entities but evolving and even perishable lineages, it still enjoys great esteem among philosophers. For instance, Putnam (1975. p. 240) affirmed that 'the essence of lemonhood is having the genetic code of a lemon', and more recently Oderberg (2007) goes back to a kind of literal Aristotelianism in equating 'essence' with 'form'.

When Putnam equates the specific essence of an organism with its genetic code ('genome' would be the proper term), the question of species re-creation is resolved. Yes, extinct species can be re-created as long as biotechnologists can handle their genetic information. Even more far-fetched thought experiments support essentialist thinking. Imagine that an extra-terrestrial entity lands on earth and looks like a dog, behaves like a dog, possess the genome of a dog and reproduces like a dog. Why should it not be called a dog even if it has no common ancestor with our terrestrial dogs? If an alien dog possess the essence of a dog, surely a cloned thylacine is a real thylacine?

Aristotelianism has almost lost all serious support among biologists and philosophers of biology. Indeed, there is no empirical way, whether morphological, genetic or other, to define a set of non-ambiguous properties equivalent to an essence. Moreover, as the Darwinian theory of evolution emphasizes change and variety in the species itself, it seems radically at odds with essentialism. As Elliott Sober (1994, p. 163) says: 'essentialism about species is today a dead issue' – at least as far as real essentialism is concerned, as we will see below.

Species as individuals

The triumphant neo-Darwinian synthesis paved the way in the 1960s and the 1970s for a new metaphysical definition of species, the so-called species-as-individual thesis. As each member of a species is unique and participates in a unique contingent process leading to the evolution of the species to which it belongs, David Hull (1994) among others defended the ontological position that species should be understood as spatio-temporally limited entities (that is individuals) whose material parts are equated with the organisms they are composed of. If the species can be called an individual in a purely metaphysical sense, it remains nonetheless very different from an individual organism which is, from a biological point of view, highly integrated, organized, homeostatic, with a determinate developmental pathway from birth to death, and many other features.

According to this metaphysical stance, when a species goes phyletically extinct (succumbing to terminal extinction), one can make a straightforward analogy with the death of an organism. It ceases to exist both functionally, as there are no more vital relations (reproductive, ecological and so on), and even materially, as no spatio-temporal entity that was part of the species exists anymore. This is comparable to a dead organism which does not exhibit any vital physiological relation amongst its internal parts (such as organs or cells) and whose organs fall increasingly apart with time.

All attempts to resurrect it from a cell or from the genetic information taken from the dead organism is doomed to failure, as this would create a new organism, that is a new spatio-temporally delimited individual, although one very similar in many aspects to the dead organism. To take a dramatic example:

> Hitler's clone, an individual possessing the same genetic information as Hitler, would never be the same individual, and far less the same person, because he would not be entirely determined by his

> genotype, but also by epigenetic effects and by the environment. Similarly, because the ecological environment of a recreated species will be different from that of the original (especially if the lapse between extinction and recreation is long), the species will become another species, that is, it will be different from what it would have been if it had evolved naturally. (Delord 2007, p. 660)

From an organism's perspective, cloning does not promise eternal life, as popular science articles or novels would have us believe.

However, a question remains unanswered. If the cloned organism does not belong to the previously *forever* extinct species (even if in its genetic content it is absolutely similar to a once alive individual of the dead species), to which species does it belong? Taking for granted that every organism must belong to one and only one species, how to decide? Fully adhering to the spatio-temporal unity definition of a species in this metaphysical frame, one has to recognize that the cloning process involves two kinds of prerequisites besides the technical skills of the scientists: on the one hand an abstract and informational part (the genetic information) and on the other hand a material part, the enucleated cell with all the cellular machinery (ribosomes, mitochondria, etc.) as well as a suitable environment for the development of the embryo (an egg or a uterus). This last part uniquely counts as spatio-temporally defined and linked without discontinuity to an organism of the surrogate species. Therefore, the new cloned organism should be linked to the surrogate species. The fact that it will certainly look and behave differently than the surrogate species implies that the biologist or the taxonomist will have to make a choice from amongst the four following options that exclude the affiliation to the extinct species E. The first option is that the clone belongs to the surrogate species S notwithstanding its distinct features (an option that appears very unlikely from a taxonomic perspective). According to the second option it belongs to a new species because its new characteristics ensure that it irremediably opens a distinctive evolutionary pathway. According to the third option it is considered as a hybrid organism (S x E). Indeed, sterile hybrids such as mules are the only organisms in the phylogenetic classification that are not assigned to any species taxon. In the botanical nomenclature, they are named by reference to both parent species, for example *Sempervivum grandiflorum* x *montanum*; but, according to the zoological nomenclature, these organisms are not even covered, and thus not labelled (ICZN, 2013). The fourth option is that these organisms can be considered as chimeras, such as grafted plants that are designated with the names of the taxonomic units composing it, joined by a '+' sign; for example *Cytisus purpureus*+*Laburnum anagyroides*. Which choice is the most suitable in the case of resurrected or de-extinct species? As a philosopher, I prefer to stay neutral on this issue. If biologists remain faithful to the species-as-individual metaphysical hypothesis, it will be their duty to adopt collectively the best (or the least inconvenient) solution. Another and ultimate option is to change their metaphysical view about the nature of species.

The perdurantist-eternalist metaphysics of species as individuals

One of the drawbacks of this first metaphysical stance is that it does not fulfil some basic requirements of a theory reflecting the evolution of species through time. For instance, how can a species remain the same and change indefinitely through time? The main issue of this metaphysics is that it is built on an endurantist-presentist conception of change (in a three-dimensional space with an extraneous time variable). As Rieppel (2009, p. 35) states:

> an enduring object is an object that is wholly present at any point in space and time at which it exists. It is impossible for an enduring object to be composed of numerically different parts in different space-time regions... This view of enduring objects is usually coupled with a presentist account of time... Presentism limits ontological commitment to the existence of objects to the present time.

The problem of this endurantist-presentist account of temporal objects is that present-time exists as a succession of infinitely small instants without any possibility of giving sense to the notion of identity over a plurality of instants when the parts of the object change from one instant to the next.

To remedy such an awkward shortcoming, a less intuitive but more satisfying metaphysics of change has been introduced in the debate about species, coming directly from the philosophy of fundamental physical objects (Hales and Johnson 2003). Taking the perdurantist-eternalist stance, Brogaard (2004, p. 226) asserts that:

> eternalism takes time as the fourth dimension that is on a par with the three dimensions of space. Space–time forms a four-dimensional continuum... On that account, all the space–time slices of an object, past, present, and future, 'co-exist' in four dimensions. A perduring

> object then forms a space–time worm, as do species on a perdurantist-eternalist account.

The addition of time as a fourth dimension is extremely convenient and parsimonious for resolving such a well-known paradox of change as Theseus's boat. According to the story, all the planks of the original boat were replaced, allowing the boat to perdure for centuries. An endurantist must reply negatively to the question whether the boat still is Theseus's boat, because all the parts of the boat are not present at the same time in the spatial form designating the boat. According to the perdurantist-eternalist approach, the boat still is the same boat, even though all of its parts have been replaced, because it remains the same space–time worm understood as a mereological sum of temporal parts (Reydon 2008).

On this account, individual species conceived as spatio-temporally limited beings cannot be really extinct; they are only 'far away' (Rieppel 2009). And as long as the newly created clone instantiates the previous spatial niche occupied by the former extinct species, it becomes instantly part of the space–time worm of the previously extinct species. Whether or not it is composed of the same biomolecules as the ancient species has no relevance here. The special feature of this worm is its absolute flatness along a certain length of its temporal dimension (which can also be interpreted as its existence in another possible world).

This ontological interpretation of species change is quite far from the commonsense view. Nonetheless, Rieppel (2009, p. 38) reminds us that in the history of biology, similar accounts of species resurrection can be found, although not under the evolutionary paradigm but under the doctrine of pre-existence:

> Resurrection is a re-newed process of unfolding. For Bonnet (1762), the development of an embryo was nothing but an unfolding of pre-existent structures, a process of evolution... all forms of life pre-existed since their initial creation, encapsulated within one another in such a way that 'new' forms of life would unfold from 'germs of resurrection', encapsulated in their 'ancestors' and perfectly adapted to the new environmental conditions that emerged after successive catastrophic events.

This perdurantist-eternalist metaphysics of species change solves beautifully and with little effort the paradox of species resurrection by biotechnological means, as well as by keeping untouched the species-as-individual ontology that many biologists consider as having no relevant alternative under the neo-Darwinian paradigm. Its constitutive weakness is its lack of intuitive appeal and the confusing debates it provokes among philosophers of biology. Reydon (2008), for instance, proposes two distinct denotations for the term 'species': as events and as objects. Only the second is considered in a four-dimensional space–time. According to Rieppel (2009), the space–time worm model cannot fully take into account the open dimension of the processes that shape the species in the future, because evolution is contingent and unpredictable. This position leads him to formulate a supplement to the perdurantist metaphysics that he calls 'futuralism'.

I fear that these confusing quibbles will do more harm than good to the legitimacy of the eternalist account of species change. But whatever the detailed formulation, I would like to insist on the fact that the question of change in the species-as-individual ontology is much better dealt with by the four-dimensional individual metaphysics than by its three-dimensional counterpart. Nevertheless, if the species-as-individual stance equates species with 'evolution in the making', or 'species as process' to quote Rieppel (2009), the alternative at looking at species as a product of evolutionary forces is not without interest, especially for taxonomists, whose classificatory skills are essential for resolving our question of the potential resurrection of biotech species.

Weak essentialism and the re-creation of species

Since Ernst Mayr (1982) strongly opposed essentialist or 'typological' thinking in the name of the synthetic theory of evolution by advocating the notion of 'population thinking', the insistence on biological variations as the vital fuel for the evolutionary motor has almost totally discredited the alternative essentialist option. The classic argument states that the intra-specific variability with respect to all organismic traits is fundamental to the Darwinian explanation of organic diversity. As essences presuppose the lack of any variation between members of the same kind, this concept sounds completely adverse to the contemporary theory of evolution.

Moreover, what could count as one of the last arguments for essentialism, and which is invoked by Hilary Putnam (1975) due to a certain level of biological ignorance or naïvety, namely the existence of a shared genetic pattern and information by all the members of a species, is clearly dismissed by biological facts. At the chromosomal and genetic levels, species taxa are distinguished by clusters of covarying traits, not by shared essences. Therefore, no definition of the real essence of a species is possible, as this would have to be based on an intrinsic property of the species members, and which was shared by them alone.

Samir Okasha, however, does not abandon the project of defining species as biological kinds or classes by presupposing that a species-specific essence can also be understood in a weaker form, as relational properties between the members of the species:

> On all modern species concepts (except the phenetic), the property in virtue of which a particular organism belongs to one species rather than another is a relational rather than an intrinsic property of that organism. On the interbreeding concept, the property is 'being able to interbreed successfully with one group of organisms and not another'; on the ecological concept the property is 'occupying a particular ecological niche'; on the phylogenetic concept the property is 'being a member of a particular segment of the genealogical nexus'. (Okasha 2002, p. 201)

Okasha then makes a distinction between two features of essentialism. According to him, the essences play both a semantic role and a causal-explanatory role in the traditional essentialist story, that is, an essence both indicates an underlying property of a class that is shared by all its members and identifies this property (this 'something else') with the causal reason why this class possesses these properties in particular. But, remarks Okasha (ibid., p. 203):

> there is no *a priori* reason why the same thing should play both of these roles. It is perfectly possible that the extension of a kind term should be determined not by superficial characteristics but by 'something else', just as Kripke and Putnam say, without it being true that that 'something else' causally explains the presence of the superficial characteristics.

Okasha then explains in the case of species how the semantic role of essences can be distinguished from its causal role, and how the former is instantiated by relational properties at the species level:

> we treat the ability to interbreed successfully as the determinant of conspecificity and we use morphological similarity as a fallible indicator of that ability. Now clearly, the causal explanation of why an organism has the particular morphological traits it does will cite its

> genotype and its developmental environment, not its ability to interbreed with certain other organisms... So interbreeding/phylogeny only play the semantic role that Kripke and Putnam attribute to 'hidden structure', not the causal/explanatory role. (Ibid., p. 204)

The same story applies again to the concept of ecological species. The ability to exploit a given ecological niche under the same environmental constraints, favouring the same adaptations, will not explain causally why it exhibits certain morphological traits and not others. Instead, this explanation requires a mix of genetic, developmental and environmental causes.

Thus, relational essentialism asserts that the relations amongst organisms determine their belonging to a given species and thus the essence of the species: their morphological or genetic features are only clues or indicators as to a presumption that they belong to the same species. One of the surprising consequences of Okasha's position is that 'two molecule-for-molecule identical organisms could in principle be members of different species, on all of these species concepts'. This statement is especially counter-intuitive since, as I mentioned earlier, for real essentialism an extraterrestrial horse, if exactly similar to a terrestrial one, would be considered as belonging to the species *Equus caballus* by virtue of its 'horseity'. Okasha's statement is nevertheless perfectly valid since the essence of the species is not determined by the intrinsic properties of its members, but only by the relational properties they exhibit at the species level. The identification of an organism to its species is accidental relatively to the inner nature or the causal structure of this organism. Henceforth, according to Okasha's relational essentialism, a cloned organism does not automatically belong to a dead species simply because it contains the genome of a past organism of this species, even if it is a perfect copy of the latter. As a matter of fact, we have to consider the relations between the cloned organisms and their past models. Despite the distance in time between the original and the cloned specimen of the extinct species, many relations still hold and make reference to the essence of the species: as long as the cloning technique is not considered as a speciation event, the re-created organisms participate in a genealogical nexus specific to their species, phylogenetically speaking. In the same line of reasoning, one can assume that the relational essence of the species is kept unchanged by cloning if this essence is grounded in infertility potential or in relatively ecological features (provided that the ecological niche of the extinct species has been preserved or can be restored).

Another version of weak essentialism, designed to fit metaphysically with the theories and practice of contemporaneous biologists, has been advocated by Richard Boyd (1999). He developed the notion of a natural kind used in classificatory schemes for explanatory and inferential practices as a 'homeostatic property cluster'. Bluntly formulated, Boyd's theory is an adaptation to the scientific realm of Wittgenstein's notion of family resemblance, initially formulated in the context of ordinary language. Boyd claims that species are paradigmatic homeostatic property clusters, as they contribute to a wide spectrum of disciplinary requirements in biology that cross-stabilize around this concept. Moreover, 'according to this "homeostatic property cluster theory", the members of a kind share a cluster of similar properties, but no property is necessary for membership in this kind' (Nanay 2011, p. 192). This conception implies in our case that the essence of a species, in the weak sense, can be attributed to the genetic cluster resulting from its history (the reproductive as well as the ecological properties of the functioning of the species), without referring to a property shared by every member of the species, in each individual genome for instance.

A given genome in this case can be considered as a sufficient property, but not as a *necessary* condition of membership of a given species. For instance, three different organisms each having two properties AB, AC and BC, considered as sufficient, belong to a species S that clusters properties A, B and C, although none of the three properties exemplifies a necessary condition for belonging to S.

Boyd's theory gives a deep epistemological background to a proposition ('species conceptualism') that I have explained elsewhere (Delord 2007) which amounts to making the notion of 'genetic cluster' or 'genetic space' a sufficient criterion for defining species membership, *in a given ecological and developmental environment*. This genetic or genomic space encompassing all the genetic diversity of the species known so far is neither predictive nor causal at the species level, a position that echoes Okasha's semantic essentialism.

Thus, Boyd's and Okasha's weak essentialism represents a solution to the problem of ascribing full membership of a species to a cloned specimen re-created from an extinct species, and for two reasons. First of all, weak essentialism provides an essence for the species that can be revived after an interruption of the evolutionary process. This essence, which is relational and based on the semantic role played by the concept of 'essence', that is designing properties that are significant from an evolutionary perspective, guarantees that we deal with the same species before and after the extinction event. Secondly, from the point of view of weak essentialism, the relational essence is neither the description of a static and immutable essence, nor a process. Ontologically, I would qualify it as an intermediate category, a cluster of relations that links both tenets of the resurrection paradox. Therefore, the concept of relational essence – though not sufficient to explain causally the transition from a product to a process of evolution, since human intention (*telos*) is required here – helps us, however, to resolve this Orphic challenge: to bring an extinct species back to life from the underworld.

Conclusion

The biotechnological manipulation (palaeogenomics, synthetic biology and cloning techniques) of the genetic remains of extinct animals opens the way to a multitude of projects, from the most scientifically informed to the most fanciful ones, so as to resurrect animals and plants from recently extinct species.

Besides all the technical and engineering difficulties such an innovative enterprise might bring about, many conceptual questions must be answered. I have indicated that beyond the need for precision that one can expect from vague notions such as 'species' or 'extinction', one has to make sure that no contradiction would preclude the authentic resurrection of the extinct species. In particular, a solution must be provided that will resolve or sidestep the resurrection paradox which states that the transformation of an evolutionary product into an evolutionary process is a real conundrum and that membership of the extinct species is not annihilated by the process of cloning.

I have examined four different metaphysical positions regarding the nature of species and have tried to assess the compatibility of them in the light of the intention of resurrecting extinct species. The two mainstream positions, real essentialism and three-dimensional individualism, fail to intrinsically deny the possibility of the recreation of species, either for biological or for logical reasons. I then analysed two alternative positions that give an original account of species change through processes of evolution and human modifications. Four-dimensional individualism refutes philosophical arguments that could banish the resurrection of species due to its imaginative dimension and its parsimonious way of reconciling the evolutionary paradigm with the reductionist and essentialist conception of the handling by biotechnologists of full genomes. Weak essentialism, whether relational or based on the homeostatic cluster concept of species, redefines the concept of 'essence' so as to be compatible with human intervention and species evolution.

Besides the fact that these metaphysical constructions are far from evident, their adoption by philosophers, biologists and biotech engineers remains problematic. Beyond their positive contribution to the resolution of the resurrection paradox, their consequences on other branches of biology and technology should be carefully assessed in order to balance the advantages and the disadvantages of such new ideas.

These four metaphysical options allow us seriously to consider the ethical dimension of the potential resurrection of species. Although this question was not the subject of this chapter, to conclude I will outline two emerging problems of (true or false) de-extinction. Indeed, whether the clone of an extinct species is considered the same as the past species or not, it could be considered as a temporal alien or as an artefact, both solutions asking for a close ethical scrutiny.

If the re-created organisms are not allowed full membership of the extinct species, then they constitute new, artificial organisms like GMOs, with a higher degree of hybridity. Despite their potential interest (for science, for education, for fun, for the environment), a strange feeling hints at their having a less-than-animal status (as poor in anima, or poor in having 'proper' life) and which is compensated for by a machine-like or artefactual ontological dimension. Thus, they would deserve less attention and care than natural beings and fall under the rules of an ethic of responsibility, where the designer of an artefact is responsible for its negative consequences (Larrère 2006).

If the re-created organisms are considered natural because they belong to a naturally evolved (though previously extinct) species, they should be submitted to the jurisdiction of an ethics of respect (respect for nature, for evolution, for life) (Taylor 1986). Nonetheless, this resurrected species becomes a temporal alien as there is a gap between the original niche and its contemporary equivalent. The acceptability of this species must be checked; it has to be monitored and controlled because it might become a source of ecological disorder and bring chaos to our temporal indigenous species, such as in the form of a geographically alien species.

More generally, given the technical and ethical uncertainties surrounding the project of species re-creation, is it worth the time, the money and the effort? If nobody remains unaffected by the positive goals promoted by this innovative 'biotech ecology' (in terms of ecosystem functioning and humanity's achievements), the final criterion should be the interests of the extinct species (if they exist!) and the interests of the biosphere as a whole, but this is another debate.

References

Bentley, R.A., M.W. Hahn and S.J. Shennan (2004) 'Random Drift and Culture Change', *Proceedings of the Royal Society. B: Biological Sciences*, 271, 1443–50.

Bonnet, C. (1762) *Considérations sur les corps organisés* (Amsterdam: Michel Rey).

Boyd, R. (1999) 'Homeostasis, Species and Higher Taxa' in R.A. Wilson (ed.), *Species: New Interdisciplinary Essays* (Cambridge, MA: MIT Press), 141–86.

Brogaard, B. (2004) 'Species as Individuals', *Biology and Philosophy*, 19, 223–42.

Danchin, É., et al. (2011) 'Beyond DNA: Integrating Inclusive Inheritance into an Extended Theory of Evolution', *Nature Reviews Genetics*, 12, 475–86.

Delord, J. (2007) 'The Nature of Extinction', *Studies in History and Philosophy of Biological and Biomedical Sciences*, 38, 656–67.

Fletcher, A.L. (2008) 'Bring 'Em Back Alive: Taming the Tasmanian Tiger Cloning Project', *Technology in Society*, 30, 194–201.

Gibson, D.G., et al. (2010) 'Creation of a Bacterial Cell Controlled by a Chemically Synthesised Genome', *Science*, 329 (5987), 52–6.

Hales, S.D. and T. A. Johnson (2003) 'Endurantism, Perdurantism and Special Relativity', *Philosophical Quarterly*, 53, 524–39.

Hull, D.L. (1994) 'A Matter of Individuality' in E. Sober (ed.), *Conceptual Issues in Evolutionary Biology* (Cambridge, MA: MIT Press), 193–215.

International Commission on Zoological Nomenclature (ICZN) (2013) 'What Organisms Does the ICZN Cover?', http://iczn.org/content/what-organisms-does-iczn-cover (accessed 30 May 2013).

Larrère, R. (2006) 'Une éthique pour les êtres hybrides – De la dissémination d'Agrostis au drame de Lucifer', *Multitudes*, 24, 63–73.

Mayr, E. (1982) *The Growth of Biological Thought* (Cambridge, MA: Harvard University Press).

Miller, W., et al. (2008) 'Sequencing the Nuclear Genome of the Extinct Woolly Mammoth', *Nature*, 456, 387–90.

Miller, W., et al. (2009) 'The Mitochondrial Genome Sequence of the Tasmanian Tiger (*Thylacinus Cynocephalus*)', *Genome Research*, 19, 213–20.

Nanay, B. (2011) 'Three Ways of Resisting Essentialism about Natural Kinds' in J. K. Campbell and M.H. Slater (eds), *Carving Nature at its Joints: Topics in Contemporary Philosophy, Vol. 8* (Cambridge, MA: MIT Press).

Noonan, J.P., et al. (2005) 'Genomic Sequencing of Pleistocene Cave Bears', *Science*, 309 (5734): 597–9.

Oderberg, D.S. (2007) *Real Essentialism* (New York/London: Routledge).

Okasha, S. (2002) 'Darwinian Metaphysics: Species and the Question of Essentialism', *Synthese*, 131, 191–213.

Orlando, L. (2005) *L'anti-Jurassic Park : Faire parler l'ADN fossile* (Paris: Belin).

Putnam, H. (1975) *Mind, Language and Reality* (Cambridge: Cambridge University Press).

Reydon, T. (2008) 'Species in Three and Four Dimensions', *Synthese*, 164, 161–84.

Rieppel, O. (2009) 'Species as a process', *Acta Biotheoretica*, 57, 33–49.
Sober, E. (1994) 'Evolution, Population Thinking and Essentialism', in E. Sober (ed.), *Conceptual Issues in Evolutionary Biology*, 2nd edn (Cambridge, MA: MIT Press), 161–89.
Stamos, D.N. (2003) *The Species Problem: Biological Species, Ontology, and the Metaphysics of Biology* (Lanham, MD: Lexington Books).
Taylor, P.W. (1986) *Respect for Nature: A Theory of Environmental Ethics* (Princeton: Princeton University Press).
Van Valen, L. (1973) 'A New Evolutionary Law', *Evolutionary Theory*, 1, 1–30.

2
The Restorationist Argument for Extinction Reversal[1]

Derek Turner

Introduction

In 2008, researchers affiliated with the Mammoth Genome Project at Pennsylvania State University announced that they had successfully sequenced about 70 per cent of the genome of the extinct woolly mammoth (Miller et al. 2008). Other scientists have been working on the genomes of the cave bear (Noonan et al. 2005), Neanderthals (Green 2006) and the thylacine, or Tasmanian wolf (Miller et al. 2009). Some scientists (for example Leonard 2008) have expressed the hope that this new work in palaeogenomics might contribute some insights to conservation genetics. Most of the public discussion, however, has focused on the prospects for reversing past extinctions.

Cross-species cloning is no longer science fiction. In 2001, Noah the gaur – the world's first cross-species clone – lived for two days before dying of dysentery. He was created using nuclear DNA from an endangered gaur and an egg cell provided by a domesticated cow. A cow was also used as a surrogate mother (Lanza et al. 2000; Vogel 2001). More recently, scientists cloned African wildcats using egg cells from domesticated cats, which also served as the surrogate mothers. The resulting clones went on to have kittens of their own (Gomez et al. 2004). Meanwhile, scientists in Japan have succeeded in cloning a mouse using a cell nucleus that had been frozen for 16 years (Wakayama et al. 2008).

Genetic engineering using DNA from extinct species is also possible. Scientists have, for example, recreated mammoth haemoglobin (Campbell et al. 2010). They identified the relevant mammoth genes and spliced them into a bacterial genome. The bacteria then produced the mammoth blood protein. And another recent study has established that thylacine DNA is still functional (Pask et al. 2008).

In March 2013, at a TEDx conference sponsored by the National Geographic Society, Michael Archer, a scientist at the University of New South Wales, announced that his team had used somatic cell nuclear transfer to create an early stage embryo of a southern gastric brooding frog, which went extinct in the 1980s. The scientists used nuclear DNA from the frogs that had been frozen for 40 years, and they transferred that nuclear DNA into the eggs of living frogs. (As of the time of writing, this work has not yet appeared in a peer-reviewed publication.)

Reintroducing an extinct population into the wild is still a matter of science fiction at this point. Nevertheless, these rapid developments show that what seems like mere science fiction at one moment can become doable within the span of just a few years. There has been quite a lot of discussion in the popular press of the ethics of extinction reversal, but environmental philosophers have been slow to bring their expertise to bear. As a result, there has been very little in the way of careful examination of the relevant ethical arguments.

In this chapter, I suggest that it makes sense to approach the issue of de-extinction from the direction of restoration ecology. In the next section I develop and elaborate one argument that I think is promising. This restorationist argument lends prima facie support to extinction reversal in some cases, and also serves as a helpful framework for thinking about which species would be better or worse candidates for de-extinction. It provides only prima facie justification, because other considerations having to do with animal welfare and resource allocation can (and, in my view, probably should) override it. I then contrast the restorationist argument with two other bad arguments for de-extinction, and show why the former provides a better framework for discussing the issue. I then address two different objections against the restorationist argument: the worry expressed by some philosophers that ecosystem health is an unhelpful metaphor, and a worry that the ability to reverse extinction would make us complacent about biodiversity loss.

The overall take-home message is that it would be helpful to begin thinking of de-extinction technology (if it ever pans out) as an additional tool that could be used in the service of thoughtful ecological restoration. Even if the goal of bringing back extinct species remains elusive, thinking about de-extinction is a useful hypothetical exercise. For those of us who believe that ecological restoration is often a very good thing, thinking about de-extinction can help us to get a bit clearer about what we're committed to.

The restorationist argument

In 2013, two legal scholars from Stanford University's Center for Law and Biosciences, Jacob Sherkow and Henry Greely, published a short essay in *Science* on the ethics of extinction reversal. In that paper, they survey a number of arguments for and against using biotechnology to reverse extinctions, and they come to the rather modest conclusions that (a) de-extinction should not receive (much) public funding; (b) de-extinction should not be banned; and (c) de-extinction should be regulated, especially because the techniques involved raise questions about animal welfare (Sherkow and Greely 2013, p. 33). In what follows, I examine one of the arguments for extinction reversal that they discuss, which I will call the restorationist argument. According to that argument, biotechnological extinction reversal could, in some cases, be part of a thoughtful ecological restoration project, where the goal of such restoration is to promote ecosystem health. This argument lends considerable prima facie support to extinction reversal in some cases.

Sherkow and Greely present the restorationist argument in the following way:

> Some researchers argue that 're-wilding' with existing species, locally extinct in particular habitats, can help restore extinct or threatened ecosystems. The same can be argued about the restoration of extinct species. The revival of the woolly mammoth as a major grazing animal in the Arctic, for example, might provide benefits by helping to restore an arctic steppe in place of the less ecologically rich tundra. (Ibid.)

Unfortunately, this formulation contains two inessential pieces that contribute nothing to the argument and may even make it seem weaker than it really is.

The first inessential feature of Sherkow and Greely's formulation of the restorationist argument is the reference to the woolly mammoth as a candidate species. There has indeed been a great deal of speculation about using biotechnology to bring back this animal (Salsberg 2000; Nicholls 2008), but it is just one of many candidate species for de-extinction, and in the end it might not be a very good one. For example, as Nicholls points out, the reproductive biology of elephants would make cloning those animals very difficult. And although some have suggested that it might be possible to restore the steppe ecosystem that mammoths occupied (Zimov 2005), the project of re-creating the

Pleistocene arctic steppe might seem overly ambitious. It might make more sense to focus on a case where a species has recently gone missing from an ecosystem that is otherwise largely intact.

The second distracting feature of Sherkow and Greely's version of the restorationist argument is their conflation of restoration with rewilding. Donlan (2005) floated the idea of 'rewilding' parts of western North America by introducing elephants, big cats and other close evolutionary relatives of the megafauna that lived there during the Pleistocene, over 10,000 years ago (cf. Martin 2005). The relationship between rewilding and ecological restoration is not entirely clear. Wildness is an environmental value quite distinct from ecosystem health, ecological integrity or biological diversity, and it's not clear that ecological restoration needs to aim at making nature wilder. Indeed, the idea that human intervention could make nature wilder seems self-contradictory. One of Donlan's original goals was to provoke restoration ecologists to think a little more clearly about historical reference conditions. Many restoration projects in North America aim to put things back the way they were prior to European settlement. Why not go further and try to (re-)create ecosystems that approximate to what was here before any humans arrived at all? Whatever one thinks about Donlan's proposal, proponents of ecological restoration need not go so far. It would have been sufficient for Sherkow and Greely to note that ecological restoration projects may involve the introduction of locally extinct species or their close relatives.

If we strip away the inessential parts of the argument as Sherkow and Greely present it, we get something like the following:

P1. In general, it is a good thing to try to promote ecosystem health.
P2. In some cases, the loss of some particular species is damaging to ecosystem health, and the reintroduction of that species would help restore the system to health.
C1. Therefore, in those cases, it is a good thing to try to reintroduce species that have gone extinct.

I'll refer to this stripped-down argument as the 'restorationist argument', though the 'reintroduction argument' would also be an apt term. This is the sort of argument that would justify reintroducing a locally extinct species as part of a run-of-the-mill ecological restoration project. For example, if we wanted to take a cornfield in the Midwestern US and restore a bit of tallgrass prairie, that might require reintroducing plant species that had not lived in the area for a very long time. Most biologists would probably judge that the relatively species-rich prairie ecosystem is healthier than the cornfield, with just one dominant species maintained by chemical fertilizers and pesticides.

Promoting ecosystem health is not the only possible goal of ecological restoration. In actual practice, restoration projects may have multiple goals at the same time. In addition to promoting ecosystem health, one might also hope to restore ecosystems to some earlier historical condition (see for example Higgs 2003, ch. 4). According to this picture, ecological restoration is more like restoring an historic house or building: the goal is historical fidelity rather than health. It is possible to imagine cases in which the goals of health and historical fidelity pull in different directions, though in many cases they will coincide. Restoration could have a number of other cultural or aesthetic aims besides. For example, Light (2000) suggests that one goal ought to be to restore our relationship with non-human nature.

In order to keep things manageable, I propose to focus rather narrowly on the goal of promoting ecosystem health (though I'll address some concerns about this notion below). One could surely also generate an argument in favour of reintroduction by appealing to the importance of historical fidelity or to some other shared value. With that in mind, the restorationist argument as formulated here represents just one strand of restorationist thinking. Still, it is an especially central strand.

Some might think that promoting ecosystem health (or perhaps ecological integrity) is not really so central to restoration ecology. William Jordan, for example, emphasizes historical fidelity over ecosystem health. In most cases, re-creating past conditions will also mean restoring ecosystems to health, but Jordan adds that 'restoration may, in certain instances, involve injuring an ecosystem' (2003, p. 23). As an example, he mentions an attempt to restore the site of a Nazi concentration camp by re-creating a barren, 'ecologically sterile' field (ibid.). In this case, the aim was to restore an ecosystem to an earlier, damaged condition. The important thing to see, however, is that even someone who, like Jordan, denies that promoting ecosystem health is the proper aim of restoration ecology can still endorse P1 of the restorationist argument. P1 makes no claim to the effect that promoting ecosystem health is *the* goal of restoration ecology. It just says that promoting ecosystem health is a good thing. The restorationist argument would lend no support to the project of restoring a barren field at the site of a concentration camp, but that does not mean that the argument is not a good one.

As formulated above, the restorationist argument contains no explicit reference to de-extinction. But the following supplemental argument takes us to a conclusion about extinction reversal.

P3. The restorationist argument lends strong prima facie justification to local extinction reversal in some cases.
P4. The difference between local and global extinction makes no difference to ecosystem health.
C2. Therefore, the restorationist argument lends strong prima facie justification to global extinction reversal in some cases.

P3 just restates the earlier assessment of the restorationist argument. Note that reintroducing a native species that has been lost is the same thing as reversing a local extinction. We often talk about a species as having gone extinct in some region or in some portion of its historic range. P4 is doing all the heavy lifting here. Why should we accept it?

Consider, by way of an example, the reintroduction of wolves into the Greater Yellowstone ecosystem. In the western US, grey wolves were locally extinct in the region by the 1970s. Prey species, such as mule deer and elk, increased in numbers with significant negative impacts on the ecosystem. In 1995, the US National Park Service reintroduced the wolves into Yellowstone.[2] The ecological impacts from the loss of the wolves in the region would have been the same even if the wolves had gone extinct completely. (Happily, they did not.) In other words, whether the wolves go extinct globally or locally makes no real difference to the health of the ecosystem from which they have been removed, because the difference between global and local extinction is just a matter of whether wolf populations persist elsewhere. Putting the same point loosely and metaphorically: the greater Yellowstone ecosystem just doesn't care whether wolves exist in other distant regions. All it cares about, so to speak, is whether *it* contains wolves.

There could be other differences between local and global extinction (or between local and global extinction reversal) that do make a difference to our all-things-considered judgement about the ethics of de-extinction in particular cases. For example, reversing a global extinction could cost a great deal more than reversing a local one, and the high cost of developing the relevant biotechnology might mean that reversing a global extinction is not 'worth it' at the end of the day. P4 just says that if concern for ecosystem health lends prima facie support to reversing local extinctions, then it also lends precisely the same degree of prima facie support to reversing global extinctions. Anyone who supports local reintroductions on the basis of the restorationist argument should also be disposed to support de-extinction in at least some cases.

At best, the restorationist argument would only provide prima facie justification for reintroducing an extinct species. That is because other considerations could well trump our concern for ecosystem health. For example, if it turned out that the only available methods for reintroducing the lost species involved causing significant suffering and distress to individual animals, concern for animal welfare could override our concern for ecosystem health. As noted above, concerns about resource allocation could also trump the restorationist argument. If the reintroduction were extremely expensive, and if it required expensive management on an ongoing basis, then we might decide on the basis of a cost–benefit analysis that the proposed reintroduction is not the best way to make use of limited resources. With these two caveats in mind (and there could be others), it is best to think of the restorationist argument as the first step in the process of evaluating a proposed reintroduction. It is the argument that motivates and justifies the reintroduction in the first place, but it may not always win the day.

Someone could hold, without any inconsistency, that the restorationist argument is a good one, while at the same time thinking that de-extinction is probably a bad idea in most cases. (Indeed, that is my view, although I will not try to defend it fully here.) In what follows, however, I will focus on the restorationist argument in isolation from these other potentially overriding considerations. It could also turn out that there are scarcely any cases in which using biotechnology to recreate an extinct species would actually do much for ecosystem health. The point is just that if we could show that, in some particular case, de-extinction could contribute to ecosystem health – a big 'if' – then we would have a significant argument in favour of de-extinction.

One concern that is sometimes raised about de-extinction is that reintroducing extinct species could actually be bad for ecosystem health:

> If the species either is released or escapes into the general environment, it might do substantial damage. Even extinct species that were not pests in their past environments could be today... Even in the same location, the passenger pigeon would today be an alien, and potentially invasive, species – perhaps another starling or even an avian kudzu. (Sherkow and Greely 2013, p. 32)

It is important to note that this legitimate concern is already built into the restorationist argument; and so it is not an objection to that argument at all. If, in some particular case, we had evidence that reintroducing an extinct species would be bad for ecosystem health, then the restorationist argument would provide no justification for the reintroduction in that case. The concern that Sherkow and Greely articulate here is just the flip side of the restorationist argument. If we have evidence that a reintroduction would diminish ecosystem health, then that would give us a strong reason not to do it.

Bad arguments for de-extinction

Sherkow and Greely (2013) survey arguments for and against de-extinction without assessing them or developing them in any detail. They take a good argument – the restorationist argument – and make it seem weaker than it really is by hitching it to the woolly mammoth and by conflating restoration with rewilding. Then they bury it amongst other arguments that are basically non-starters. I want to consider two of those bad arguments for de-extinction here because doing so will help to highlight some of the advantages of the restorationist argument.

The first weak argument for extinction reversal appeals to the knowledge we might gain from studying, say, a re-created woolly mammoth:

> De-extinction could allow scientists the unique opportunity to study living members of previously extinct species (or, at least, close approximations to those species), providing insights into their functioning and evolution. (Ibid., p. 33)

One major problem with this argument is that you can learn a great deal about woolly mammoths, say, without going all the way through with the de-extinction. The research team at Penn State that sequenced the woolly mammoth genome was initially and ostensibly seeking answers to questions about mammoth evolution. For example, did the mammoths living in Asia and those in North America constitute two genetically distinct sub-populations? The scientists who re-created mammoth haemoglobin were interested in finding out whether mammoth blood would have been effective at transporting oxygen at low temperatures. It is possible to investigate these questions without recreating a mammoth. If anything, the hype about de-extinction has drawn attention away from this interesting, pure scientific work.

Furthermore, and sticking with the mammoth example for a moment, it's not entirely clear what we could learn about the woolly mammoths of the Pleistocene by re-creating a herd of them using, say, cross-species cloning. Although the newly created animals would all presumably have mammoth nuclear DNA, their mitochondrial DNA and other cellular machinery would all be derived from Asian elephant egg donors. The first ones would all be born to elephant surrogate mothers and raised in captivity in an environment completely different from that of their Pleistocene predecessors. If the idea is to learn something about, say, mammoth social behaviour, it just isn't clear that de-extinction, even if it were wildly successful, would enable scientists to draw any solid inferences. The fact that the newly created animals would only be 'approximations' of extinct creatures from the past significantly weakens the argument from scientific knowledge. And in the cases of some recently extinct species, such as the thylacine, we already have quite a lot of information about their behaviour and ecology from observations made in the not-so-distant past.

However, the fact that de-extinction would only give us 'approximations' of extinct plants or animals is not such a problem for the restorationist argument. In cases where a species has gone extinct locally, it could be that the best that restorationists can do is to introduce a closely related population from a different region. The fact that the newly introduced population is not exactly the same as what was there before may not be such a problem if there is good evidence that introducing these natural approximations would be good for ecosystem health.

This argument concerning scientific knowledge would be far more convincing if one could point to some specific question of empirical historical science that could only be answered by going through with a de-extinction. This line of argument could conceivably lend more justification to some related lines of research. For example, although using cloning or genetic engineering to re-create a dinosaur is not on the cards, palaeontologist Jack Horner has been promoting a line of research that involves intervening in the development of chicken embryos to try to create birds with teeth and other dinosaurian traits (Horner and Gorman 2009). The aim is not so much to re-create some animal that lived during the Mesozoic but simply to learn more about evolution and development.

Sherkow and Greely seem to place more weight on a second argument, which one can only call the 'argument from coolness':

> The last benefit might be called 'wonder' or, more colloquially, 'coolness'. This may be the biggest attraction, and possibly the biggest benefit, of de-extinction. It would surely be very cool to see a living woolly mammoth. And while this is rarely viewed as a substantial benefit, much of what we do as individuals – even many aspects of science – we do because it's 'cool'. (Sherkow and Greely 2013, p. 33)

I agree that it would be cool to see a living woolly mammoth. But this argument is underdeveloped at best. The argument could just as easily count against de-extinction as for it. (Incidentally, it shares this defect with the worn out 'playing God' objection.) We might, after all, come to think that it would be cool to decide *not* to use biotechnology to bring back extinct species. Many philosophical questions need answering before we can take this argument seriously. What does it mean to say that something is cool, in the first place? Can we be mistaken about what's cool? And so on.

Sherkow and Greely are betting on a wide consensus about the coolness of seeing a living woolly mammoth, and that seems like a safe bet. Woolly mammoths are charismatic megafauna, and they loom large in popular culture. But what if we were talking about reversing the extinction of an ecologically significant plant species that few non-specialists had ever heard of? The argument from coolness would get very little traction. Moreover, in coming years it is likely that much of the discussion of de-extinction will shift away from the general question as to whether we should do it at all, focusing more on the question of which species would be better or worse candidates for it. The argument from coolness gives us no useful way of addressing that prioritization problem. The coolest species, such as the woolly mammoths, could well turn out not to be very good candidates for de-extinction, perhaps because animal welfare concerns are more serious, or because it would take such a massive restoration effort to re-create the ecosystems in which the animals lived. By contrast, the restorationist argument does give us a way of approaching the prioritization problem. Start by identifying ecosystems that are most in need of restoration, and then determine whether reintroduction of lost species would be helpful in those particular cases. Focus on places, rather than species.

The argument from scientific knowledge and the argument from coolness are not too promising. The restorationist argument avoids the problems that confront these two. Nevertheless, there are some potential problems with the restorationist argument, and I will turn now to discuss two of those.

Objections

Ecosystem health is just a metaphor, and not a very good one

The premise that does much of the work in the restorationist argument is P1, the claim that we should try to promote ecosystem health. Some philosophers have expressed doubts about the value of this notion (Jamieson 1995; Russow 1995). Jamieson and Russow raise a number of different worries, and I will not be able to address all of them here. So for now, the restorationist argument should receive at best a qualified endorsement, pending a more complete assessment of the concept of ecosystem health. Nevertheless, I will address a couple of Jamieson's concerns, and I will suggest that the concept of ecosystem health is no more problematic than the ordinary concept of human health. Although I will not undertake this project here, the concept of ecosystem health needs to be re-evaluated in the light of recent developments in medicine and microbial ecology.

The concept of ecosystem health is a metaphorical one. That itself is not necessarily a strike against it, since many of the central concepts of ecology and conservation biology are metaphorical (Larson 2011). Think of invasive species, biological communities or ecological niches. Nor is environmental science at all distinctive in this regard. Many, perhaps most, scientific concepts have metaphorical origins: random genetic drift, DNA transcription, chemical bonds, electromagnetic waves, the big bang, phylogenetic trees, evolutionary arms races, etc. When you start looking for metaphor in science, you soon begin to see poetry everywhere. Indeed, in an opinion piece on extinction reversal, science writer Olivia Judson (2008) took to calling it 'resurrection science', a metaphorical term with obvious religious resonances. We also routinely use metaphor to think about our relationship with non-human nature (Turner 2005). For just one example, consider Aldo Leopold's famous injunction to stop thinking of ourselves as conquerors of the land and start thinking of ourselves as 'plain citizens' of the biotic community (Leopold 1949).

Jamieson (1995) worries a lot about other contexts in which people use the health metaphor for rhetorical or political purposes. For example, talk about 'healthy families' can be a way of giving the appearance of authoritativeness to normative views about families that are politically and ethically controversial. A public policy that aims to promote healthy families sounds all well and good, until we learn that advocates for the policy think that the healthiest families are heterosexual couples with children and at least one stay-at-home parent. In cases like this, the talk of health is a way of making it seem like one's prejudices have some objective basis. Jamieson is right to raise our suspicions about health talk in contexts outside of medicine. But it turns out that there is a significant difference between the notion of ecosystem health and the other examples of medical rhetoric that he is worried about. The concept of ecosystem health turns out to be extremely important in medicine itself.

Increasingly, medical researchers have begun to think that many human health problems have to do with our microbiota. Our health is largely a matter of the staggeringly complex interactions among the populations of microbes that inhabit (or perhaps help constitute) our bodies. Because it has only recently occurred to medical researchers to study our microbiota, it is at present difficult to say just how many and what sorts of human ailments are traceable to what might be described as ecosystemic problems. But it is becoming increasingly clear that medicine has quite a lot to do with microbial ecology. The human microbiome project has now succeeded the human genome project (Juengst and Huss 2009). Doctors are already beginning to explore medical treatments that quite literally involve the reintroduction of microbial varieties that seem to be missing from a person's system. In other words, there are cases in which ordinary medicine just *is* microbial restoration ecology. Here is one description of such a procedure:

> Recently, out of desperation as much as anything else, doctors have resorted to what seems like an extreme treatment: fecal transplants. Doctors obtain fecal bacteria from healthy donors – normally family members – and place them in the patients' intestines, usually during a colonoscopy. There have been only a few trials, but the results have been astounding. In one study, all thirty-four recipients were cured; these are people for whom all other approaches had failed. (Specter 2012, p. 38)

The patients were suffering from infections of the gut bacteria *C. difficile*. In most cases, other friendly bacteria keep *C. difficile* populations from getting out of control and causing health problems. But people sometimes experience problems when those friendly bacteria get killed off by antibiotics. The solution: reintroduce the helpful bacteria. Such cases look a lot like local extinction reversal. The sort of reasoning that might justify the reintroduction of wolves into Yellowstone is no different from the reasoning that would justify fecal transplants.

In this fascinating development, a metaphorical concept (ecosystem health) is being reapplied back to the original source domain (human health and sickness) in a way that is very likely going to change how we think about human health. This metaphorical feedback loop places critics of the notion of ecosystem health in an awkward situation. If some human health problems turn out to be failures of ecosystem health, then it will not be possible to dismiss the notion of ecosystem health so easily. Talking about healthy ecosystems turns out to be rather different than talking about healthy families, because the concept of ecosystem health matters in medicine.

Jamieson (1995) observes that people generally do not like being sick or injured. Illness and injury are unpleasant for people, and this is one reason for seeking to restore the afflicted to health. Larger scale ecosystems, by contrast, do not mind being unhealthy. They have no first-person perspectives, feel no pain or distress, and have no cares or preferences. Thus, Jamieson argues that one of the main reasons for promoting human health is absent in the case of larger ecosystems. Even Jamieson would have to acknowledge, though, that people's first-person experiences of illness and injury are not the only reason we have for treating those illnesses and injuries. Some illnesses (for example, some forms of mental illness) might not be experienced as such by the patient. In light of the foregoing argument concerning microbial ecology, we might say that we have extra reason to promote the health of our own ecosystems – that is, of ourselves. But this does not mean that we have no reason at all to promote the health of the ecosystems of which we are part. P1 of the restorationist argument is perfectly compatible with Jamieson's observation that larger ecosystems cannot mind being sick.

Some theorists, such as Higgs (2003, pp. 122–3), prefer the notion of ecological integrity to ecosystem health. However, Higgs's discussion of these conceptual issues could be clearer. To start with, it isn't clear how 'integrity' is really any different from 'heath'. Higgs writes that:

> At the very root of integrity is the notion of wholeness, which in the context of conservation and restoration suggests that the goal ought to be the creation of whole, intact ecosystems. (Ibid., p. 122)

How is a 'whole, intact ecosystem' different from a healthy one? Moreover, some of his specific concerns about the notion of ecosystem health apply with equal force to the notion of integrity. For example, he worries that 'there is so much variation in ecosystems that criteria for ascertaining health are either too broad to be practically useful, or too specific to capture a full range of meaning' (ibid., pp. 123–4). Why would this not be just as much a problem for the notion of ecological integrity? It's not clear that we gain anything much by focusing on integrity rather than health. However, those who, like Higgs, prefer the notion of ecological integrity can just replace 'health' with 'integrity' in P1 and P2 of the restorationist argument.

There is an ecocentric tradition in environmental philosophy, going back at least to Aldo Leopold (1949), which takes ecosystem health to have intrinsic value (cf. Nelson 1995). This view actually coheres quite well with the view that human health is one kind of ecosystem health. It seems plausible to say that human health has intrinsic value; but that is tantamount to saying that the health of certain sorts of microbial ecosystems has intrinsic value. This at least makes it plausible that the health of other sorts of ecosystems could have intrinsic value as well. Ecosystem health clearly also has instrumental value, since human communities depend on the functioning of the larger ecosystems in which they are embedded. Thus, ecocentrists and anthropocentrists can happily converge in defence of P1 of the restorationist argument.

Much more needs to be said about this topic, and my discussion here only scratches the surface. However, I do hope to have shown that in general one way to address worries about the notion of ecosystem health is to point out that human health is one kind of ecosystem health, and that some medical treatments can take the form of microbial restoration ecology.

De-extinction would make us complacent about biodiversity loss

So far, I have tried to show that the restorationist argument lends prima facie justification to de-extinction, provided that the concept of ecosystem health survives further philosophical scrutiny. There is, however, one other familiar objection against de-extinction that targets the restorationist argument quite specifically. We can call this the 'complacency objection':

> Current protection of endangered and threatened species owes much to the argument of irreversibility. If extinctions – particularly extinctions where tissue samples are readily available – are not forever, preservation of today's species may not seem as important. (Sherkow and Greely 2013, p. 33)

The worry here is that if extinction turns out to be reversible, then we might be more complacent about preventing extinctions in the first place. Sherkow and Greely treat that objection as a serious one and do not pause to consider how a proponent of the restorationist argument might reply.

Unfortunately, those who make the complacency objection rarely do the work of spelling it out in any detail. It is possible to construe the objection as a worry about resource allocation – which is a worry that I share. Suppose (somewhat artificially) that we have a limited amount of funding available for the promotion of ecosystem health. We have to decide whether to invest our limited funds in preventing the loss of existing biological diversity or in reversing extinctions. It could well be that investing big in extinction reversal is not the most cost-effective way of promoting ecosystem health, and that it would be smarter to allocate our limited funding to loss prevention. Moreover, where resources are limited, investing big in extinction reversal could be a sign of complacency about biodiversity loss.

There is more to the complacency objection than this legitimate concern about resource allocation. The deeper issue has to do with the difference between reversible and irreversible environmental harm. One might think that, other things being equal, it is more important to prevent irreversible damage than reversible damage. Much of the impetus for conservation biology derives from the thought that extinction is forever. If extinction turns out to be reversible, then it will just seem less important to prevent extinctions in the first place. Perhaps this is the sense in which de-extinction would make us more complacent about biodiversity loss. In the remainder of this section, I will explore this second way of understanding the complacency objection.

Sherkow and Greely do not point out that this second version of the complacency objection has already received some attention in the literature on restoration ecology. In his classic paper, 'Faking Nature', Robert Elliot (1982) engages with a version of the complacency objection. Elliot is worried about cases in which corporations, especially those involved in extraction, promise ahead of time to repair or restore any environmental damage that their activities will cause. His response to this concern is to argue that the environmental damage is not, in the end, reversible, because of the way in which causal history contributes to value. Even if the company hired restoration ecologists who (miraculously) put things back exactly the way they were, some value would have been lost because the restored ecosystem would not have the right sort of causal history. In support of the claim that history contributes to value, Elliot invites us to consider the difference in value between a great painting – say a masterpiece by Rembrandt – and a qualitatively identical fake produced by a team of super-forgers. The fake will have less value because it has the wrong sort of history, even if we assume that it is utterly indistinguishable from the original. For this reason, restoration ecologists will never be able to put nature back the way it was. In the very best case scenario, their efforts will produce a 'fake' that is indistinguishable from what was there before, but which has less value because it does not have the right sort of causal history. This reasoning can serve as the basis for a response to the complacency objection: we should not be complacent about ecosystem damage, because that damage is always irreversible in at least this one respect.

Elliot's reply to the complacency objection may not be the most effective one, however. In order to see why, imagine a healthy ecosystem in which one important species – say, wolves – goes locally extinct. This local extinction leads to a noticeable decline in ecosystem health. To address the problem, conservationists reintroduce a new population from elsewhere. That new population has a different causal history than the original one. Human conservationists and wildlife biologists figure prominently in the causal history of the newly reintroduced population. Presumably, however, humans had played no such role in the history of the original population. Perhaps Elliot would argue that some value has been lost because the newly reintroduced population has the wrong sort of history. From the perspective of ecosystem health, however, the difference between the history of the original population and the history of the reintroduced population should make no difference. To put the point very loosely: how the wolves got there makes no difference to ecosystem health. What matters, from the perspective of ecosystem health, is how many wolves there are, what sorts of things they eat, how much they eat, which territories they establish, and so on. Even if damage to ecosystems involves some irreversible loss of value (which is Elliot's point), the damage to ecosystem health could well be largely reversible.

There is another, better response to the complacency objection. It is possible to get some additional mileage here out of the medical analogy. Suppose that there is a very serious disease for which we (so far) have no effective treatment. The disease is irreversible. We know how to prevent the spread of the disease, but the preventive measures are rather burdensome and require significant effort and sacrifice. So far, public health professionals have focused most of their efforts on getting people to adopt those preventive measures. Now, however, a team of researchers announces that they are very close to finding a safe and effective cure that would render the disease reversible. Suppose that a bioethicist responds to this news by arguing that researchers should not work to develop a cure, because the public health efforts aimed at preventing the spread of this disease rely on the claim that it is irreversible. If it turned out to be reversible, then people would become more complacent. The disease would not be so bad. When we transpose it into a medical context, the complacency objection seems to fall flat. Of course scientists should work on developing a cure for the disease in question, even if having a cure would make people more complacent about prevention.

To see more clearly where the complacency objection (on the second construal) goes wrong, it will help to distinguish between the descriptive and normative issues. On the one hand, there is the descriptive psychological claim that people are generally more complacent about reversible damages, and hence that having the ability to reverse extinctions would make people more complacent about biodiversity loss. Then there is the normative question of whether it is rational to be more complacent about reversible damages. Suppose for a moment (as seems plausible) that it's rational to be more complacent about reversible damage. In that case, if extinction reversal were to become technologically feasible, it would be rational to be more complacent about preventing extinction. So it is hard to see why it would be a problem if we became more complacent about biodiversity loss. Now suppose that it's not rational to be more complacent about reversible damage, and that we should take care to prevent damage without regard to its reversibility. In that case, the reversibility of extinction would not be a reason to care less about preventing it. Having the ability to reverse extinctions should (if we are rational) make no difference at all to how much we care about preventing extinction in the first place. To the extent that we are rational, de-extinction would not make us more complacent about biodiversity loss. Either way, it's very difficult to see just where the problem is supposed to lie.

To sum up the results of this section: the complacency objection might well be a good one if we construe it as an argument about resource allocation. If, however, we read it as a more principled objection against finding ways to reverse damage that we previously thought was irreversible, the objection is very difficult to make out.

There is one related conceptual issue having to do with the meaning of 'extinction'. It could be an analytic truth that extinction is irreversible. In that case, if scientists did succeed in recreating, say, a woolly mammoth, then it might be best to say that the animal had never really gone extinct at all. We might want to distinguish quasi-extinction from genuine extinction, and say that the woolly mammoths underwent a quasi-extinction 10,000–12,000 years ago. On this view, 'extinction reversal' is an oxymoron. However, it is not at all unusual for the meanings of scientific terms to evolve under pressure from technological changes, and it would hardly be surprising if the meaning of 'extinction' changes as scientists work on developing extinction reversal technology. The issue here, however, is not whether extinction is reversible, but whether certain kinds of biodiversity loss are reversible. It does not matter much for present purposes whether we call those losses extinctions or quasi-extinctions.

Conclusion

Anyone who thinks that the restorationist argument justifies the reintroduction of locally extinct species should also allow that it would provide prima facie justification for full blown extinction reversal in some cases – namely, in those cases where bringing back an extinct species would help to restore a damaged ecosystem to health (or if you prefer, you might substitute the notion of ecological integrity). Although concerns about animal welfare and resource allocation should temper our enthusiasm for de-extinction and might well trump the restorationist argument, that argument is a good one. Philosophers' concerns about ecosystem health can be addressed to some degree by pointing out that human health is largely a kind of microbial ecological health. The worry that the capability to reverse extinctions would make us complacent about biodiversity loss is also misguided. We are still in the very early stages of discussion of the ethics of extinction reversal, but we would do well, going forward, to think of de-extinction in the context of restoration ecology.

Notes

1. Many thanks to the other participants at the workshop on re-creating extinct species held at the University of Turku, Finland, for their helpful feedback on an earlier version of this chapter. Thanks especially to Markku Oksanen and Helena Siipi for their thoughtful comments. I presented versions of this chapter at Connecticut College and at an ISEE session at the Pacific Division meeting of the American Philosophical Association, and am grateful to those audiences for their comments and questions. This chapter has benefitted a great deal from conversations with colleagues (especially Simon Feldman) as

well as students in my Philosophy of Biology and Environmental Philosophy courses at Connecticut College.
2. For more information, see www.nps.gov/yell/naturescience/wolfrest.htm (accessed 25 May 2013).

References

Campbell, K.L., et al. (2010) 'Substitutions in Woolly Mammoth Hemoglobin Confer Biochemical Properties Adaptive for Cold Tolerance', *Nature Genetics* 42(6), 536–9.

Donlan, J. (2005) 'Re-Wilding North America', *Nature*, 436, 913–14.

Elliot, R. (1982) 'Faking Nature', *Inquiry*, 25(1), 81–93.

Gomez, M.C., et al. (2004) 'Birth of African Wildcat Cloned Kittens Born from Domestic Cats', *Cloning and Stem Cells*, 6(3), 247–58.

Green, R.E. (2006) 'Analysis of One Million Pairs of Neanderthal DNA', *Nature*, 444, 330–6.

Higgs, E. (2003) *Nature by Design: People, Natural Processes, and Ecological Restoration* (Cambridge, MA: MIT Press).

Horner, J. and J.Gorman (2009) *How to Build a Dinosaur* (New York: Penguin).

Jamieson, D. (1995) 'Ecosystem Health: Some Preventive Medicine', *Environmental Values*, 4(4), 333–44.

Jordan, W.R., III (2003) *The Sunflower Forest: Ecological Restoration and the New Communion with Nature* (Berkeley, CA: University of California Press).

Judson, O. (2008) 'Resurrection Science', *The New York Times*, 25 November 2008, http://opinionator.blogs.nytimes.com/2008/11/25/resurrection-science/ (accessed 29 January 2011).

Juengst, E. and J. Huss (2009) 'From Metagenomics to the Metagenome: Conceptual Change and the Rhetoric of Translational Genomic Research', *Genomics, Society, and Policy*, 5(3), 1–19.

Lanza, R.P., et al. (2000) 'Cloning of an Endangered Species (*Bos Gaurus*) Using Interspecies Nuclear Transfer', *Cloning*, 2(2), 79–90.

Larson, B. (2011) *Metaphors for Environmental Sustainability* (New Haven, CT: Yale University Press).

Leonard, J.A. (2008) 'Ancient DNA Applications for Wildlife Conservation', *Molecular Ecology*, 17, 4186–96.

Leopold, A. (1949) *A Sand County Almanac* (Oxford: Oxford University Press).

Light, A. (2000) 'Ecological Restoration and the Culture of Nature: A Pragmatic Perspective' in P.H. Gobster and R.B. Hull (eds) *Restoring Nature* (Washington, DC: Island Press), 49–70.

Martin, P. (2005) *Twilight of the Mammoths: Ice age Extinctions and the Rewilding of America* (Berkeley, CA: University of California Press).

Miller, W., et al. (2008) 'Sequencing the Nuclear Genome of the Extinct Woolly Mammoth', *Nature*, 456(20), 387–91.

Miller, W., et al. (2009) 'The Mitochondrial Genome Sequence of the Tasmanian Tiger (*Thylacinus Cynocephalus*)', *Genome Research*, 19, 213–20.

Nelson, J.L. (1995) 'Health and Disease as "Thick" Concepts in Ecosystemic Contexts', *Environmental Values*, 4(4), 311–22.

Nicholls, H. (2008) 'Let's Make a Mammoth', *Nature*, 456(20), 310–14.

Noonan, J.P., et al. (2005) 'Genomic Sequencing of Pleistocene Cave Bears', *Science*, 309(5734), 597–600.

Pask, A.J., et al. (2008) 'Resurrection of DNA Function in vivo from an Extinct Genome', *PLoS One* 3(5), e2240, www.plosone.org/article/info:doi%2F10.1371%2Fjournal.pone.0002240 (accessed 20 February 2011).

Russow, L.-M. (1995) 'Ecosystem Health: An Objective Evaluation?', *Environmental Values*, 4(4), 363–9.

Salsberg, C.A. (2000) 'Resurrecting the Woolly Mammoth: Science, Law, Ethics, Politics, and Religion', *Stanford Technology Law Review*, 1, 1–30.

Sherkow, J.S. and H.T. Greely (2013) 'What if Extinction Is Not Forever?', *Science*, 340(6128), 32–3.

Specter, M. (2012) 'Germs Are Us', *The New Yorker*, 22 October 2012, 32–9.

Stone, R. (2013) 'Fluttering from the Ashes', *Science*, 340(6128), 19.

Turner, D. (2005) 'Are We at War with Nature?' *Environmental Values*, 14, 21–36.

Vogel, G. (2001) 'Cloned Gaur a Short-Lived Success', *Science*, 291(5503), 409.

Wakayama, S., et al. (2008) 'Production of Healthy Cloned Mice from Bodies Frozen at −20°C for 16 years', *PNAS* 105(45), 17318–22.

Zimov, S.A. (2005) 'Pleistocene Park: Return of the Mammoth's Ecosystem', *Science*, 308, 796–98.

3

What's So Special about Reconstructing a Mammoth? Ethics of Breeding and Biotechnology in Re-creating Extinct Species[1]

Christian Gamborg

Introduction

Wild animals are objects of fascination and concern. Efforts are made all over the world to protect wild animals – especially the ones in danger of losing their habitats or their life – and even more so when it is the last member of a species or subspecies. Although difficult to assess, an educated guess places the current extinction rate to about a dozen species a day (Chivian and Bernstein 2008), although there is a great deal of uncertainty about this figure (May 2011). Species losses have their greatest effect when the species lost was previously abundant or had a so-called functionally irreplaceable role in the given ecosystem (Wardle et al. 2011). Certainly, a sense of urgency would seem to form part of the empirical basis for placing the science and practice of de-extinction well into the domain of environmental conservation. In addition, there is an alleged moral imperative because humans in many, but not all, cases have (in)directly caused the species to disappear (Wolverton 2010). Hence, 'repair' of ecological damage and species extinction is seen as 'an act of enlightened self-interest, as well as an ethical imperative' (Cairns 2003). It can also be seen as a more pleasure-seeking reason: that by restoring or rewilding damaged ecosystems we may bring wonder back into our lives (Monbiot 2013). But is it for real? 'Mammoths Soon to Come Alive', 'Breeding Ancient Cattle back from Extinction' (Faris 2010) or 'Clone Zone: Bringing Extinct Animals Back from the Dead'

(Frozen Ark 2011) – headlines like this have the ring of slightly dated science fiction or the scrapped titles of a somewhat low key sequel to the hit movie *Jurassic Park*, now with the sabre-toothed tiger and the cave bear from the Pleistocene epoch (which spans the most recent glaciations). But there is increasingly less science fiction about the on-going projects around the world – as reconstruction technologies develop – that involve trying to resurrect: the Pyrenean Ibex, a kind of wild mountain goat known as a bucardo which was officially declared extinct in 2000; the thylacine, a hyena-like marsupial carnivore, which vanished as a species in the 1930s; the passenger-pigeon which went extinct in 1914; and several more – not least the woolly mammoth which disappeared during the Holocene warming. A race has started with some scientists claiming that they believe it is just a matter of years before it will be possible to produce a healthy, cloned mammoth by using tissue from the body of a four-month old mammoth, preserved by permafrost (Switek 2013a).

Resurrection efforts are not just about ways of changing animals but about ways of changing nature. The re-creation of extinct species seems to steer right into a long-standing ethical debate about the (right) use of technology, about what constitutes a fair distribution of harms and benefits, and about the nature and extent of our responsibilities. This debate concerns on the one hand the moral status of (resurrected) animals and issues about how they should be treated; and on the other hand the value of nature and what kind of meddling is considered acceptable. One of the stated reasons for taking up the challenge and financially supporting this kind of research endeavour is one of getting more knowledge about extinct species. Another often stated, and more applied, purpose is that it could help to preserve nature by re-creating extinct species and indirectly, by using the knowledge gained from doing so, especially with regard to techniques, to help saving endangered species. Following an over-utilization and extermination of species, which was followed by comprehensive conservation and preservation efforts, as expressed by the Convention on Biological Diversity, a main nature management trend in Europe and in North America has been ecological restoration, such as habitat or landscape restoration and including species reintroduction. But not only is a change taking place in the actual practices and stated justifications as substantial losses of species are witnessed (Hoffman et al. 2010), there is also a move towards 'breaking up' old dichotomies between conservation and preservation, between active and passive nature management. Clearly, if it is only feasible to produce one individual these goals cannot be reached and the products of these efforts will only find homes in zoos, cabinets of rarities or museums. Nevertheless, if it was possible to produce a population it could be argued that it might, in theory, be seen as making a difference in restoring lost nature.

In this the chapter I will explore whether there is anything special about reconstructing a mammoth – that is, the issue of re-creating species – from two different, albeit related, angles. The first looks at the ethical aspects of specific techniques (breeding and modern biotechnology) to bring about such extinct species, particularly the use of cloning. This kind of technology gives rise to ethical questions. Are breeding-induced animal welfare problems acceptable? What ethical limits are in play here, and how should they be elaborated given the plurality of ethical perspectives? The other angle places species re-creation in the context of nature restoration, specifically relating it to the practice of back breeding, asking whether such radically reintroduced animals should be regarded as research animals, animals in our care or as wild animals. I will also consider how such restoration should be conceived, whether as 'real' nature or nature by proxy.

Ethical issues related to techniques of bringing extinct species back to life

History tells us that advances in science are difficult to predict. Current pursuits in the name of re-creating species are based on selective breeding and other kinds of reproductive technologies, and cloning and genetic engineering. Animal breeding was until the beginning of the twentieth century a relatively uncontrolled activity based mainly on the animal's physical appearance. Domesticated animals descend from wild animals as their wild ancestors gradually started to live in close proximity to humans and were later, for generations, bred under human control. These animals were not only tamed but have been adapted genetically over many generations.

Domestication leads to dramatic changes in the physical appearance of an animal, as can be observed in the differences between a wolf and the number of dog breeds witnessed today. Domestication is also associated with behavioural changes with most domestic animals being calmer and less fearful than their wild ancestors (Sandøe and Christiansen 2008). Typically it also brings about changes in reproductive biology. Thus, while their wild ancestors display strict seasonal reproduction rhythms, most domesticated species can reproduce all through the year. The early animal breeders did not really have the knowledge and tools to predict and control what they were doing. This changed in the first half of the twentieth century when Mendelian genetics was applied. The second half of the twentieth century saw the development of new forms of animal biotechnology – all of which allow scientists and breeders even greater control over future animals. Modern breeding practices and various reproductive technologies have delivered significant results across a wide range of applications, a range which now has become even wider with the potential ability to re-create extinct species.

Reproductive technologies aim to control, and often accelerate, the process of breeding. The first technology of this kind to be developed was artificial insemination (AI). This allowed reproduction to take place without natural mating. In the 1950s a technique for freezing semen ensured that AI would become even more significant since it could now be stored over a longer period of time and transported to a geographically much wider area. Similarly technologies have been developed to enable female animals to produce more progeny than they would naturally, such as super-ovulation and embryo transfer, the latter making it possible to shuttle embryos to surrogate mothers. A related technique is one which makes it possible to remove immature eggs from female animals, mature and fertilize them *in vitro*, and then transfer them to recipient animals which serve as surrogate mothers, relevant for re-creating extinct species. One of the more spectacular forms of biotechnology – paramount to the quest for trying to resurrect species – has been the type of animal cloning which really got started with the sheep Dolly, born in 1996. Dolly originated from a cell taken from the udder of her biological mother. The result was the first mammal to be cloned from an adult animal. Since then cat, dog, horse, pig, camel, mouse, fish and monkey, to name a few, have been cloned by somatic cell nuclear transfer where DNA is removed from an unfertilized egg and replaced with DNA from the cell of an adult organism. After the embryo has been grown in culture, it is inserted into the surrogate animal's uterus.

So far, the main interest in cloning has come from scientists involved in basic or biomedical research. Basic research may be designed to improve our understanding of embryonic development, or deepen our knowledge of epigenetic processes. In biomedical research the interest lies in using reproductive cloning as a tool enabling efficient production of animals that can serve as disease models. In doing this, cloning is combined with another form of modern biotechnology: genetic modification (also referred to as genetic engineering). It involves the direct manipulation of an organism's genetic make-up by introducing new genes to a fertilized egg or it involves the 'knocking-out' of specific genes so they no longer function. It is also possible to move genes across species barriers (known as transgenesis).

But a new interest in the use of such technologies has come up: re-creating extinct species. In this new branch of conservation biology, cloning – alone or in tandem with other kinds of reproductive technology – could form part of a future conservation strategy for very endangered species such as: the banteng, an ox from South East Asia; the Sumatran tiger; or the giant Panda. Cloning could also be used to bring back extinct species like the woolly mammoth, the Pyrenean ibex or the Tasmanian tiger. (If a close descendant exists, back breeding is an option – which will be discussed in the next section.)

There are many pathways for the use of cloning and genetic engineering. Eggs can be fertilized *in vitro* using frozen sperm – if the target species (for example, an elephant) and the egg donor (for example, a mammoth) are similar enough to produce a viable embryo. In that case, a hybrid would be produced – which theoretically could then be back-crossed. Another approach is taking out the DNA of a surrogate egg and injecting it with a copy of the extinct species' genome obtained through well preserved somatic cells (as in the case of the Pyrenean ibex). In this case, a pure clone would be obtained. If there is no intact genome, it must be created from scratch, provided it is known how the DNA is organized into chromosomes. This is no small feat as the largest genome to have been synthesized is still a thousand times smaller than the mammoth's. If part of the DNA exists it may be possible to reconstruct the whole molecule.

At present all of these avenues of trying to re-create species are being pursued – but not without problems. With deep freeze resurrection, the challenge is obtaining frozen sperm from the long-dead animal (such as the mammoth). In the second approach, troubles mount in the reassembling of DNA from tissue when the molecule is too fragmented. Using the example of the woolly mammoth, its nearest relative, the elephant, has a genome which is complete, and over 85 per cent of the mammoth's genome has been sequenced – which would suggest a way forward that involved engineering a synthetic version or altering the elephant's genome in the relevant places, that is modifying the elephant chromosomes where they differ. Then, in principle, the genes for specific mammoth features like big tusks could be added to the elephant genome, inserted in an embryo and then implanted in an elephant surrogate.

There are obstacles when it comes to using cloning for resurrecting species. One is finding suitable and intact DNA. Another is the inefficiency of the process where it is only possible to implant a few embryos. Moreover, many cloned animals suffer from impaired health, including placental abnormalities, foetal overgrowth, prolonged gestation, stillbirth, hypoxia, respiratory failure and circulatory problems, malfunctions in the urogentical tract, malformations in the liver and brain, immune dysfunction, anaemia, and bacterial and viral infections. In the case of the cloning of the first endangered species – the ox-like gaur from South East Asia – a cow was used to bring the cloned baby gaur to term – but it died two days afterwards from a bacterial infection. In the case of the cloning of the Pyrenean ibex – the first animal born from an extinct subspecies – only 57 eggs (out of a total of 439 used to form embryos) were implanted into surrogate goats. Of these 57, only five went through the entire gestation period, and only one was born alive (Folch et al. 2009). This uncertainty about the outcome, the 'spending' of animal material, the comprehensive manipulation and the very real health problems have contributed to provoking negative reactions from the public (Gamborg et al. 2009).

Public perception and ethical concerns

Although our knowledge of public perception of animal cloning in the context of species re-creation is limited, owing to a lack of systematic qualitative and quantitative studies focusing on this specific issue, we can try to make some general observations based on other applications of modern biotechnology such as cloning. A first observation is that cloning always exists in a context. Only at the basic research level might it be considered a stand-alone technique. Cloning is, in other words, to a large extent an enabling technology mainly used in connection with other technologies, especially those involved in genetic modification, and operating within the context of its application (Lassen et al. 2006). Thus, when it comes to public perception, cloning is often placed within the context of other (animal) biotechnologies and applications. In this view, it is possible to see a difference between especially medical and agricultural applications, with medical application being more positively received. Lassen (2005, p. 17) looked at public responses and developed two scales in relation to them: one was the organism involved; the other was the type of application: 'on the first of these scales, cloning sits towards the controversial end, since its object is animals. On the second scale the position depends on the purpose and application of the cloning being considered'. Thus, considering both scales, an application like cloning for food production would seem controversial. With regard to cloning for re-creating animal species, and the possible use of genetic modification, on that scale it would most likely also be considered controversial. When it comes to the context of application, it would probably be less controversial than the use of cloning for, say, food purposes. Having said that, public perception is difficult to gauge since it depends not only on an absolute acceptance/rejection of the technology in question but to a large degree on the perceived usefulness and existence of *alternatives*.

Several critics have pointed out that efforts should rather go into 'traditional' conservation efforts like habitat restoration – as molecular 'gimmickry simply does not address these core problems' (Pimm 2013). It is also argued that species revival would not curb the activities that decimated the animals in the first place, would target only a very few species, and would be what Ehrenfeld (2013) has dubbed 'just an interesting idea, what we might call recreational conservation'. Indeed, one might also find the exercise of re-creating pointless, since, when a species is lost, from an *ecosystem* view, the niche it filled and the 'functions' it had will in many cases be occupied by other species (Wardle et al. 2011). Hence, the assessment of alternatives also depends on the view taken: do we focus on the individual species or the relations it has with the ecology? Moreover, from a risk reducing perspective, re-creation of species via cloning does not take away the dangers of, say, poaching, climate change and other factors which may have been contributing to the species going extinct (Switek 2013b).

Clearly, many concerns can be, and have been, raised in relation to re-creating species. Re-creation of extinct species seems to give rise to the same type of dilemmas that have been seen in regard to the various forms of selective breeding in relation to other uses of the techniques of cloning. These moral choices are complicated because we are not only to decide how we treat current living animals but what future animals are going to exist, and what they are going to be like. The concerns include: *animals* (for example with regard to animal welfare, as high levels of early deaths and deformities are witnessed; or animal integrity, as manipulation may be seen as taking place at a fundamental level); *humans* (for example biosafety or 'slippery slope' concerns, that is the anticipated fear that animal cloning may pave the way for reproductive human cloning); and *society* (understood, for example, as a concern for loss of biodiversity or as a needless exploitation of animals for human purposes). Another widely felt concern relates to the limits of science (for example the claim that scientists are not able to predict the consequences of what they are doing). An often suggested way to address this concern is that limits should be placed on scientists' interference with nature, though this needs further investigation as there may be two claims or views at play. It can be claimed that if we try to manipulate nature on the basis of 'grand plans' for the future, there is a real danger that things may go badly wrong – as indeed it sometimes has when species of plants and animals have been introduced by humans into new territory and therefore may bring about some kind of disaster; and likewise if a resurrected animal turns into a pest (invasive species) or acts as a vector for pathogens. Alternatively, the concern may be an expression of a view that we should leave nature as it is because untouched nature has a value of its own, that we should respect what is seen as the integrity of nature, or that we are making choices which are not for us to make. How do these thoughts fit into a more comprehensive account of our duties to animals? Answers to these questions will clearly depend on the ethical perspective applied – which will be sketched in the following.

From a 'model' contractarian perspective, potential and actual benefits (such as joy or awe of mere existence or, more concretely, for nature restoration purposes) speak in favour of using breeding and modern biotechnology. This would call for moderation if short-term benefits were being achieved at the price of significant damage in the longer term. For example, there might be good reasons to be cautious about bringing back extinct species to life if they carry diseases, become a pest or somehow have a negative effect on the specific person holding this view. It might also matter from a contractarian perspective if a large proportion of the general public were concerned about the application of biotechnology in re-creating extinct species – not out of concern for the animals but because that use of biotechnology may alienate or disgust people which in turn may then create problems for one self.

From a 'model' utilitarian perspective, the welfare of the affected animal become parts of the moral judgement that there is nothing inherently problematic about engineering animals – that is by conventional breeding methods, cloning or reconstruction of DNA – as long as this is done with a view to maintaining or enhancing their welfare. It is, in this perspective, legitimate to kill an animal if this animal is replaced by another that will lead to at least as good a life. Similarly, it is legitimate to change animals through breeding and biotechnology if the animals that come out of this process are at least as well off as the animals that would have come into existence had breeding and biotechnology not been applied. In the case of re-creating extinct species, the latter point is a bit more complicated as one could argue that the alternative here, without the application of biotechnology, is not another animal, but no animal. However, the key dilemma is balancing human need, interest or preference with the cost to the animals, such as if animals are bred in such a way that they will suffer more seriously from 'resurrection' related diseases. However, if the application is considered the best means of increasing welfare, including the welfare of humans (which, it could be argued, could be seen as improved via the species coming into re-existence), then, from the utilitarian perspective, it is not only acceptable but our duty to breed animals and use biotechnology in the process. On the issue of the so-called integrity of nature, utilitarians will presumably agree that there is nothing inherently problematic about interfering with nature. The only concern here may be about being blinded by technological optimism that places too much focus on the potential benefits of breeding and biotechnology and too little attention on the more serious consequences. Looking back at previous experiences, such as the case of the Pyrenean ibex, the first clone born alive which died shortly after due to lung defects, the utilitarian may well note the negative side-effects of both breeding and biotechnology on animal health and so adopt an attitude of caution.

From a 'model' animal rights perspective, application of breeding and biotechnology to re-create extinct species is inherently troublesome. The rationale of this ethical perspective is to protect those animals that happen to exist – they have rights (Regan 2007). This means that there are certain things that we may not do to these individuals – even it means foregoing a greater good. That is, it would not be permissible to sacrifice an existing cat with a minor health problem for a future cat with a better health. In the case of re-creating species, however, most of the choices connected with the application of breeding methods and biotechnology are not between an existing and a potential animal. They are, as explained previously, between different potential animals. In some cases though, defending animal rights may mean a clear, typically negative, stance on breeding and the application of biotechnology to animals. This is because the rights view takes an opposing stance to some of the activities of which breeding and biotechnology are typically part of. Thus according to the animal rights view the use of animals in experimentation, as is the case currently with species re-creation, and technical or species knowledge should be gained in other ways, as should conservation, it could be argued. From an animal rights point of view, extinction may be seen as a disaster – but mainly directly from an environmental standpoint for the few or last remaining members of that species as habitats disappear or for the suffering of other individual animals linked to or dependent upon animals from that species, as the notion of 'species' is an abstract, and rights are assigned to individuals or collectives of individuals.

None of these ethical perspectives seem to capture concerns about not interfering with nature. From a perspective of respect for nature what matters ethically is not only to look after the interests and rights of individuals. Larger entities – or what one might term abstract entities – such as ecosystems or species also matter in out ethical reasoning. We should from this perspective look after and protect what is considered original nature. We should certainly not try to shape everything around us according to our own interests and plans (Rolston 1989). From a respect for nature point of view it may be claimed that nature as it happens to have evolved is revered and thus a very cautious and restrictive policy should be encouraged when it comes to animal breeding and biotechnology in the service of engineering the re-creation of extinct species, especially as it deals with (potential) wild animals. The question arises as to what status resurrected animals should have – are they wild or 'natural' in any sense, or, because they are manipulated, more to be regarded as a variant of domesticated animals?

Ethical issues related to fitting re-creation of species within a framework of nature restoration

So far in this chapter, emphasis has been on the ethical implications of the techniques aimed at re-creating species. Now we turn to one potential aim of re-creating species seen in the context of nature restoration, particularly when the species – in an ecosystem perspective – had a functionally irreplaceable role. The argument here is that, for example, the resurrection of a woolly mammoth may help to restore the arctic steppe from the tundra (Zimov 2005).

Nature restoration comes in degrees. There is a range of measures from management at the landscape level to the tiniest reconstruction of DNA. We have active landscape and ecological management practices to improve habitats to make them suitable for species returning through natural migration (such as the wolf). Another measure is to move individuals of a species to a former habitat, which is part of the native range – even though we are talking of, say, two thousand years in the case of the beaver. Or, in times of climate change, individuals of species are moved to new, similar ecosystems if their own habitats are about to disappear – which is the practice of managed relocation (Minteer and Collins 2010).

Species reintroduction can be seen as a limited type of ecological restoration – a type used where a particular species is missing in an ecosystem. There is also the practice of back breeding or de-domestication (for example aurochs) in order to establish whole populations over time with the aim of reverting to a more 'wild' status in their appearance, behaviour and hardiness. This can be seen as an option when (some) genetic variation of the extinct species can be found in the descendant. Moreover – as was mentioned above in relation to technologies – restoration in terms of reintroduction may be brought about by the use of certain types of biotechnology such as cloning from a single specimen. This results, potentially, but with great difficulties, in creating a viable breeding population in sexually reproducing animals of endangered species (such as the Asian gaur). When it comes to already extinct species this may concern cloning involving the reconstruction of functional DNA (for example the Pyrenean ibex).

What is worth noting about resurrection ecology is that it is a countermovement at the very last stage in the sequence of species loss over time, as described by Wardle et al. (2011). This sequence begins with habitat loss and fragmentation, leading to declining population, with increased inbreeding, and entering a so-called extinction vortex – and in the process getting added to the International Union for the Conservation of Nature Red List. With increasing rarity comes the collapse of interactions with associated plant and animal life of a region. At the later stage animals of that particular species are only occasionally observed, and from an ecological point of view its contribution to ecosystem functioning is marginal. Eventually, the species is extinct, evidently having no further effect on community or ecosystem processes.

In the following, the focus will be on fitting re-creation of species within a framework of conservation biology by looking at the practice of back breeding. Back breeding has a real advantage compared with cloning as it creates a whole population rather than just an individual animal. Back breeding resembles de-domestication, the latter being a process by which a population of animals become de-adapted to man and captivity, and adapted (to a degree) to the wild environment from which they came, by a combination of genetic changes taking place over generations and environmentally induced events experienced in each generation (Price 1984). Apart from bred-back 'wolves' such as the Tamaskan wolfdog (which does not involve cross-breeding with wolves) and the so-called Quagga project, which seeks to bring back an extinct subspecies of zebra called Quagga, the most prominent and best examples of back breeding schemes are probably the Heck cattle – a hardy breed of cattle developed in the early twentieth century in an effort to breed back modern cattle to a presumed ancestral form, the aurochs. Back breeding aims to increase the so-called wildness and naturalness of the animal/species in question, and of an area in a long-term nature management strategy. Ideally, in time, the population should be able to reproduce and become self-sustainable.

Actual back breeding projects work in several idealized phases (see Koene and Gremmen 2002), involving behavioural and gene pool changes. Such a project begins with the introduction of the back-bred animal into the new environment, then an ability to maintain itself and form a group of self-sustainable animals which over time develop distinct social behaviour and culture. In a subsequent phase individual variation in the gene pool is noticeable, and selection pressure arises and the gene pool changes to adapt to the environment – eventually, in theory, a phase is reached with a 'natural' gene pool and 'natural' behaviour in a 'natural' environment. The last phase of back breeding automatically sets a problem, however, because the animal population will only behave like their ancestors if the genes responsible for their natural behaviour are intact and have not been lost during the previous process of breeding. In both practical and theoretical terms, back breeding can very much be regarded as 'work in progress' – and by no means an uncontroversial practice (Keulartz 1999).

Wildness, naturalness and 'real' nature

Because the exact definition and actual practice are open to interpretation, it can be asked whether this process of re-creation and rewilding by any means is really possible, or whether it is just a matter of changed breeding goals and methods. In any case, such practices raise questions of the norms governing animals and nature. Is populating the landscape with animals through re-created extinct species too much intervention? Is it merely a reasonable way to make the landscape and nature suit us? And if we restore, do we get a second-rate imitation of the real thing (Elliot 1982) or wildness by proxy? In other words, is this practice a case of re-creation or merely new creation? These questions in fact concern the moral status of the back bred animals, issues about how they should be treated and the value of nature (Gamborg et al. 2010). During the stages where wildness or naturalness is thought to 'increase', how should the animals undergoing back breeding, thereby reviving extinct animal species with their genome manipulated, be conceived: as wild or unwild animals?

In the early phases of back breeding, the animals still bear a strong resemblance to their current close relative, and maybe less to their common ancestor. On the other hand, as described in the ideal phase model, the animals should gradually become more fully adapted to different environmental conditions and thus, as a population, be more like wildlife; yet they may still not be considered wild animals in the full sense. When, if at all, should we cease to regard the animals individually (and thus to be treated in accordance with welfare legislation covering animals in our care)? The question is how these animals should be regarded, ethically speaking, since the norms covering animals in our care (such as domestic animals or research animals) are very different from the norms regarding our treatment of wildlife and wild nature (see Norton et al. 1995).

Wildness might easily be thought of as a quality, in specific individual animals, of being wild or unwild. But it may also be regarded as a broader concept and likened to parts of nature that are not controlled by humans. The lifecycles of animals that are wild in this latter sense are free of intentional human intervention. Hence, this kind of wildness cannot be preserved in human dominated environments (Jamieson 1995). Clearly, an underlying assumption of this anti-interventionist conception of wildness is that humans are in some sense 'unnatural'; an assumption that has been queried (Callicott 1994). The related concept of 'naturalness' – an equally debated term (Siipi 2008) – and much used in connection with restoration ecology, is often defined, likewise, as a quality or state of ecosystems without human interference. Somewhat paradoxically, the creation of naturalness is sometimes thought to be possible through planned 'natural' disturbances (Peterken 1996). An important aspect of naturalness here is natural behaviour. If we take the example again with the aurochs, in Europe the last ones were recorded at the beginning of the seventeenth century, but only recently have information and data been compiled and analysed to build a picture of the physical appearance and behaviour and of their supposed impact on the habitats in which they lived. When it comes to re-created species, the uncertainty is a lot bigger. It could also be questioned whether animals of the re-created species would in fact be considered natural or native to the area they formerly belonged to, as the landscape may have changed considerably since the extinction (Sherkow and Greely 2013). As such back breeding as a way to re-create extinct species, as here described, share characteristics with ecological restoration in general: suitable reference points must be found, valid data must be used to flesh the scheme out, and the present state of the environment must be compared with the conditions prevailing when the environment originally existed. Using back breeding to re-create extinct species is therefore an exercise in approximation with an unpredictable result and an end-point that is hard to define.

This likeness moves us into a long-standing discussion within environmental ethics about authenticity. This debate, originating with Elliot (1982), has subsequently been revived and expanded upon by, among others, Katz (1992), Elliot (1997), Light (2000) and Chapman (2006). The question concerns whether we should regard a re-created species as authentic, a proxy or a fake? The fakery discussion has also been framed as a clash between two paradigmatic views of nature and/or biodiversity: a so-called historical view and an end-state, or consequentialist, view (Gamborg and Sandøe 2004). This simple distinction allows us to elaborate on the value of back breeding. To say what is good and right as judged by end-state principles, we do not need any information about the way this state of affairs was brought about. On historical principles, by contrast, legitimacy or acceptability depends entirely on past developments. Here information about events of the past is not merely relevant, or interesting, but essential to the determination of moral value (ibid.). Model advocates of the historical view would take naturalness to be a goal in itself, or end value. They would attach no value to restoration schemes involving back breeding to re-create an extinct species – at least as long as there were 'true' preservation initiatives to support. By contrast, a straightforward consequentialist would judge end-state principles sufficient and would welcome restoration practices as an instrument to create more genetic and landscape-related diversity. Hence, this distinction may help to explain some of the differences in view of the acceptability or benefits of trying to bring vanished species back to life.

Conclusions

Is there anything special about reconstructing a mammoth? Undoubtedly, from a scientific point of view, the practice denotes something new. Once a species becomes extinct, it is history, literally. Now, it seems that resurrection ecology may be a new frontier – of conservation biology. In relation to reproductive biotechnology, it signifies novel ways of using cloning (in terms of application, material and techniques) and seems to push forward DNA reconstruction technology and in a way extend genomics to natural ecosystems, from simple models to complex field systems (see Whitham et al. 2008). But when considering the ethics of breeding and biotechnology in re-creating extinct species – which may be conceived of as yet another expression of our increasing control over nature, and indeed life – asking if there is something new, if there is a need for a special 'mammoethics', the answer seems less affirmative than from the purely scientific point of view. Certainly, the moral choices involved in species re-creation are complicated because we are not only to decide how we treat current living animals but what future animals are going to exist, and what they are going to be like. But that goes for all type of breeding practices – this is just another application.

When it comes to resurrection seen as part of a bigger scheme of nature restoration, using back breeding, this is a practice caught between two sets of norms governing animals and nature. As such, it stirs up a range of commonly made assumptions about current nature conservation practice and our treatment of animals within it. These ultimately concern human responsibilities to animals and the specifically human conception of nature's value. As an ethically assessable practice, re-creation of extinct species does not easily take on established principles of nature conservation or contextual standards of animal treatment. There is a need to ponder more systematically the principled answers to questions about when – or if – resurrected animals move from animals in our care to animals not in our care, that is obtain their (former) wild status, and what kinds of duty are owed to such animals. Consideration of these ethical issues may help to promote a more nuanced debate about whether, and the extent to which, this type of intervention and manipulation represented by species revival, and human interference in general, in the so-called course of nature, is desirable or indeed acceptable.

Note

1. Peter Sandøe is gratefully acknowledged for his contribution to previous co-authored work and ideas which this chapter draws upon and for useful comments to an earlier version of this chapter.

References

Cairns, J. (2003) 'Reparations for Environmental Degradation and Species Extinction: A Moral and Ethical Imperative for Human Society', *Ethics in Science and Environmental Politics*, 2003, 25–32.

Callicott, J.B. (1994) 'A Critique of and an Alternative to the Wilderness Idea', *Wild Earth*, 4, 54–9.

Chapman, R.L. (2006) 'Ecological Restoration Restored', *Environmental Values*, 15, 463–78.

Chivian, E. and A. Bernstein (eds) (2008) *Sustaining Life: How Human Health Depends on Biodiversity* (New York: Oxford University Press).

Ehrenfield, D. (2013) 'Resurrected Mammoths and Dodos? Don't Count on It', www.theguardian.com/commentisfreee/2013/mar/23/de-exinction-efforts-are-waste-of-time-money, (accessed 22 October 2013).

Elliot, R. (1982) 'Faking Nature', *Inquiry*, 25, 81–93.

Elliot, R. (1997) *Faking Nature: The Ethics of Environmental Restoration* (London and New York: Routledge).

Faris, S. (2010) *Breeding Ancient Cattle back from Extinction*, http://content.time.com/time/health/article/0,8599,1961918,00.html (accessed 13 September 2013).

Folch, J., et al. (2009) 'First Birth of an Animal from an Extinct Subspecies (Capra pyrenaica pyrenaica) by Cloning', *Theriogenology*, 71, 1026–34.

Frozen Ark (2011) 'The Frozen Ark Project Newsletter', www.frozenark.org, (accessed 13 September 2013).

Gamborg, C. and P. Sandøe (2004) 'Beavers and Biodiversity: The Ethics of Ecological Restoration' in M. Oksanen and J. Pietarinen (eds) *Philosophy and Biodiversity* (New York: Cambridge University Press), 217–37.

Gamborg, C., et al. (2009) 'The Ethics of Animal Cloning' in J. Gunning et al. (eds) *Ethics, Law & Society* (Aldershot: Ashgate), 43–54.

Gamborg, C., et al. (2010) 'De-domestication: Ethics at the Intersection of Landscape Restoration and Animal Welfare', *Environmental Values*, 19, 57–78.

Hoffmann, et al. (2010) 'The Impact of Conservation on the Status of the World's Vertebrates', *Science*, 330, 1503–9.

Jamieson. D. (1995) 'Wildlife Conservation and Individual Animal Welfare', in B.G. Norton et al. (eds) *Ethics on the Ark: Zoos, Animal Welfare and Wildlife Conservation* (Washington and London: Smithsonian Institution Press), 69–73.

Katz, E. (1992) 'The Big Lie: Human Restoration of Nature', *Research in Philosophy and Technology*, 12, 231–41.

Keulartz, J. (1999) *Struggle for Nature: A Critique of Environmental Philosophy* (Florence, KY: Routledge).

Koene, P. and B. Gremmen (2002) *Gewogen Wildheid: Samenspel van Ethologie en Ethiek bij De-domesticatie van Grote Grazers* (Wageningen: Wageningen University).

Lassen, J. (2005) 'Public Perceptions of Farm Animal Cloning in Europe', www.bioethics.dk/Egne_udgivelser/~/media/Bioethics/Dokumenter/Egne_udgivelser/Cebra_rapporter/9_Public_perceptions.ashx.

Lassen, J., et al. (2006) 'After Dolly: Ethical Limits to the Use of Biotechnology on Farm Animals', *Theriogenology*, 65, 992–1004.

Light, A. (2000) 'Ecological Restoration and the Culture of Nature: A Pragmatic Perspective' in P.H. Gobster and R.B. Hull (eds) *Restoring Nature: Perspectives from the Social Sciences and Humanities* (Washington, DC: Island), 49–70.

May, R.M. (2011) 'Why Should We be Concerned about Loss of Biodiversity?', *Comptes Rendus Biologies*, 334, 346–50.

Minteer, B.A. and J.P. Collins (2010) 'Move it or Lose It? The Ecological Ethics of Relocating Species under Climate Change', *Ecological Applications*, 20, 1801–4.

Monbiot, G. (2013) *Feral* (London: Allen Lane).

Norton, B.G., et al. (eds) (1995) *Ethics on the Ark: Zoos, Animal Welfare and Wildlife Conservation* (Washington and London: Smithsonian Institution Press).

Peterken, G.F. (1996) *Natural Woodland* (Cambridge: Cambridge University Press).

Pimm, S. (2013) *Opinion: The Case of Species Against Species Revival*, www.news.nationalgeographic.com/news/2013/03/130312–deextinction-conservation-animals-science-extinction-biodiversity-habitat-environment/ (accessed 22 october 2013).

Price, E.O. (1984) 'Behavioural Aspects of Animal Domestication', *Quarterly Review of Biology*, 59, 1–32.

Regan, T. (2007) 'The Case for Animal Rights' in H. Lafollette (ed.) *Ethics in Practice*, 3rd edn (Maldon, MA and Oxford: Blackwell), 205–11.

Rolston, H. (1989) *Environmental Ethics: Duties to and Values in the Natural World* (Philadelphia: Temple University Press).

Sandøe, P. and S.B. Christiansen (2008) *Ethics of Animal Use* (Oxford: Blackwell).

Sherkow, J.S. and H.T. Greely (2013) 'What if Extinction is not Forever?', *Science*, 340, 32–3.

Siipi, H. (2008) 'Dimensions of Naturalness', *Ethics and the Environment*, 13, 71–103.

Stone, R. (2013) 'Fluttering from the Ashes?', *Science*, 340, 19–20.

Switek, B. (2013a) 'Will We Ever See the Woolly Mammoth again? What about the Striped Tasmanian Tiger, Once-Prolific Passenger Pigeon, or the Imposing Wild Cattle Called Aurochs?', www.news.nationalgeographic.com/news/2013/13/130310-extinct-species-cloning-deexinction-genetics-science/ (accessed 13 September 2013).

Switek, B. (2013b) 'The Promise and Pitfalls of Resurrection Ecology – Phenomena: Laelaps', http://phenomena.nationalgeographic.com/2013/03/12/the-promise-and-pitfalls-of-resurrection-ecology/ (accessed 13 September 2013).

Wardle, D.A., R.D. Bardgett, R.M. Callaway and W.H. van der Putten (2011) 'Terrestial Ecosystem Responses to Species Gains and Losses', *Science*, 332, 1273–7.

Whitham, T.G., et al. (2008) 'Extending Genomics to Natural Communities and Ecosystems', *Science*, 320, 492–5.

Wolverton, S. (2010) 'The North American Pleistocene Overkill Hypothesis and the Re-wilding Debate', *Diversity and Distributions*, 16, 874–6.

Zimov, S.A. (2005) 'Pleistocene Park: Return of the Mammoth's Ecosystem', *Science*, 308, 796–8.

4
The Authenticity of Animals

Helena Siipi

Introduction

Is a passenger-pigeon-like bird that has come into existence through de-extinction procedures an authentic passenger-pigeon? A similar question can be asked about all animals produced by the three methods of de-extinction: back-breeding, cloning and genetic engineering (Sherkow and Greely 2013, p. 32). The question of authenticity also concerns endangered animals that are genetically modified to be more likely to survive in their changing environment. On what condition is a genetically modified Sumatran tiger still an authentic Sumatran tiger? These questions are important in considering the justification of de-extinction methods on extinct species and the genetic modification of endangered species. If it turns out that these animals cannot, even in theory, be authentic members of the original species, the justification for using the de-extinction methods and genetic modification for conservation and other purposes is considerably weakened.

The term 'authentic' is ambiguous. In the contexts of de-extinction and the genetic modification of endangered animals it has two main meanings: identity meaning and quality meaning. According to the identity meaning, if something is not an authentic x, then it is not x at all, but merely something else that more or less resembles x. The claim that a bird born from de-extinction procedures is not an authentic passenger-pigeon may then be interpreted to mean that the bird is not a passenger-pigeon. Rather it is a bird that resembles real passenger-pigeons. This identity meaning of 'authentic' makes a distinction between being and not-being something, in other words a distinction between a real and an unreal something. The term 'authentic' is commonly used in this sense when distinguishing an original from a copy or something real or genuine from a fake. Authenticity is then seen as an either–or distinction that corresponds to class membership: if something is not an authentic x, then it also fails to be x. With respect to de-extinction and the genetic modification of endangered animals, the identity meaning of authenticity raises questions regarding the species identity of the animals born from these procedures. Do genetically modified animals and animals born from de-extinction procedures belong to the original species? Are they proper members of it, or should we rather see them as copies or even as fakes?

According to the second main meaning, namely the quality meaning of authenticity, the claim that the produced bird is not an authentic passenger-pigeon means that the bird is a certain kind of passenger-pigeon. It is a poor, lousy and less ideal passenger-pigeon, but a passenger-pigeon nevertheless. People commonly use the term 'authentic' in this sense, for example when they judge a building or a work of art to be a more or less authentic representative of a certain style (such as rococo or art deco) or a food offer to be a more or less authentic Italian pizza, Japanese sushi or an American cheesecake. As the examples indicate, the quality meaning of authenticity concerns quality, typicality and the value of entities. Quality meaning relies heavily on the concept of ideal or perfect (and less ideal and less perfect) members of a certain class or type. Authenticity in this sense is a question of degrees. With regard to de-extinction and genetic modification of endangered animals the following questions arise. How good or ideal a Sumatran tiger is a genetically modified Sumatran tiger? Are the animals produced by genetic modification and de-extinction procedures less good, typical or valuable than the animals that have come into existence without human technological involvement?

The identity meaning of authenticity and species identity

Originality, realness and species identity

The identity meaning of authenticity is closely connected to the terms 'original' and 'real'. 'Original' refers to something being the first of its kind or an instance of something that is the first of its kind (Gilmore and Pine 2007, pp. 49, 57–8). Being an original can be contrasted with a copy or a fake. Copies may be highly similar to the originals, sometimes to the point of being indistinguishable regarding their actual physical properties. Yet, copies differ from the originals at least with respect to their histories (the way they have come into being) and are, thus, distinct from the originals and not instances of them (Gunn 1991, p. 303;

Gilmore and Pine 2007, p. 57). When a copy is falsely claimed to be an original, or an instance of the original, it becomes a fake. This intent to deceive is always present in fakes. Fakes differ from other copies in that they are pretended to be something they are not (Radford 1978, p. 76; Gunn 1991, p. 303).

The term 'real' is broader than the term 'original'. Just as in being original, being real can also be contrasted with being a copy or a fake. However, the sources of realness go beyond being the first of one's kind. In the sense of realness, authenticity is a question of being what one is claimed to be, or being true to oneself (Gilmore and Pine 2007, p. 96–7; Erler 2011, p. 238). The understanding of authenticity as realness is consistent with the idea of variations. Variations differ to some extent from the originals (the first instances of something). Contrary to copies and despite these differences, variations retain their identity as real (instances of something). Different variations of strawberries, for example, differ from original strawberries, but are nevertheless considered as real strawberries.

To answer the question about the authenticity of animals born from procedures of de-extinction and genetic modification, it needs to be found out whether they should be seen as (instances of) the original and something real, or as copies. Further possibilities include seeing them as fakes or as variations. Both de-extinction and the genetic modification of endangered animals raise these questions, but they raise them from different reasons. This is due to the different goals of these two approaches. The goal of de-extinction is to produce animals that are as similar to the animals of the species that once died out as possible. In genetic modification of endangered species the properties of animals are intentionally changed. De-extinction aims at broad similarity, whereas genetic modification aims at miniscule dissimilarity. Thus, with respect to de-extinction questions are answered by analysing the relevance of the way an entity has come into being to its realness and originality. These issues also concern genetically modified animals, but in their case the degree of sufficient similarity to the original animals also needs to be discussed.

The above questions regarding de-extinction and genetic modification connect to each other due to a common conservational aim behind them. Both aim at a future existence of a certain species. Thus, combinations of the two are possible and have been suggested. Re-created animals might, for example, be genetically modified to make them immune to the pathogen that caused their extinction in the first place (Rosen 2012; Revive & Restore 2013). Moreover, in practice the question of sufficient similarity concerns de-extinction too. Perfect clones of the original extinct animals cannot, at least in the near future, be produced (Ehrenfeld 2006, p. 730; Sherkow and Greely 2013, p. 32). As noted by the Revive & Restore project (2013), 'de-extinction projects will not produce species that are 100 per cent genetically identical to the extinct species, due to the constraints of working with incomplete ancient DNA'. Further problems result from the fact that some animals rely on parental guidance in developing species-typical behaviour patterns. Some animal species typically live in herds, and it may be questionable whether a lonesome individual that has no contact with other members of its species can develop as the individual that has social contacts with other members of its species would (Ehrenfeld 2006, p. 730). Despite these difficulties, the statement of the Revive & Restore project (2013) of the Long Now Foundation is full of hope: 'it is expected that the revived species will be nearly identical genetically, and "functionally identical" ecologically'. Even if total similarity is not required, a high degree of similarity is necessary for successful de-extinction. Accepting this, nevertheless, does leave open the questions regarding the meanings of a 'high degree', 'sufficient degree' and 'similar enough'.

The concept of extinction

Supposing that technological difficulties can be solved and animals that have a strong resemblance to the members of the original species can be brought to life, then what reasons would there be to reject the assertion that these animals are authentic and real members of the original species? At least two conceptual reasons for seeing them rather as copies (or even as fakes) can be found. The first is based on a certain understanding of extinction; the second relies on a view of species as a type of entity that cannot be re-created.

According to philosopher Alastair S. Gunn (1991, p. 299), extinction is always and necessarily final: 'extinct also says something about the future of the class – that once it becomes a null class, it can never come to have members again. It may be claimed that this is what extinct means'. Gunn is, thus, most unwilling to accept the commonly presented message of the resuscitation scientists that, due to new technological developments, 'extinction may not be forever' (see for example Rosen 2012; Redford et al. 2013, p. 3; Sherkow and Greely 2013, p. 32). For Gunn, the irreversibility of extinction does not follow, and has never followed, from technological limits or inabilities. Rather irreversibility is embedded in the concept of extinction: being final is necessary for a disappearance to be an extinction. If some disappearance is refutable, then it is not an instance of extinction. And thus, because of the so-called definitional stop, the question 'is de-extinction possible?' must always be answered negatively (Gunn 1991, p. 299).

Although this kind of understanding of extinction certainly rules out the possibility of de-extinction, its argumentative power against animal re-creation may be questioned, as it carries with itself a load of assumptions and debatable implications. The view that extinction is necessarily final implies that we have to accept one of the following three claims as well as its implications. The first option is to accept that, even though morphological, genetic and functional properties of animals born from re-creation procedures were totally similar to animals of the original species, they cannot and should not be considered as members of that species. As a result, the following problem (Savulescu 2011, p. 656) emerges: if passenger-pigeon-like birds born from de-extinction procedures are not passenger-pigeons, what are they? The question concerns the species identity of the animals. The proponents of this first alternative might argue that animals born from de-extinction procedures belong to a new human-created species (Garvey 2007, p. 151), and this new species can and should be understood as a copy of (and thus distinct from) the original species.

The second possibility is to accept that animals born from the de-extinction procedures are members of the species that once died out, and that, despite their existence, the species remains extinct. This is, of course, a highly provocative statement, acceptance of which requires an explanation of how it is possible for a species to be extinct even though animals belonging to it are alive.

A third possibility is to accept animals born from de-extinction procedures as members of the species in question and to contest the view that the extinction took place before they came into being. This third view implies that a species may fail to be extinct even though no animal belonging to it is alive.

Gunn (1991, pp. 298–301) argues for the first claim and, because he relies on the view that extinction takes place when a last individual of a species dies, omits to take seriously the third alternative. The third alternative, however, may not be as self-evidently wrong as it first seems. Philosopher Julien Delord (2007, pp. 659–60) has shown that extinction can be conceptualized in several ways that differ with respect to the point at which extinction is seen to take place. Most commonly the point of extinction is associated, as Gunn does, with the death of the last member of a species. It may, however, also be associated with the end of the last couple (end of the potentiality to produce offspring) or, as

Delord suggests (ibid., p. 661), with the loss of genetic (or other) information necessary for producing an individual with characteristics of the species. When extinction is understood in this way, the third possible implication of the irreversibility of extinction becomes sensible and understandable. Extinction may be final, but, in the case of animal recreation, it never really took place, since information regarding the species was not lost and it was possible to bring its members into existence by technical means. Irreversibility of extinction thus becomes compatible with the idea of animal re-creation – although, since extinction does not (and did not) take place at the moment of death of the last individual, animal re-creation should, then, strictly speaking, not be called 'de-extinction'.

To conclude, the idea that irreversibility is a necessary condition for extinction does not necessarily rule out the possibility of animal re-creation. The compatibility of animal re-creation and the irreversibility of extinction, nevertheless, requires accepting the view of extinction as a loss of information necessary for producing individuals of the species in question. If one wants to hold on to the traditional view of extinction taking place when the last individual of a species dies, then one has to either (as Gunn does) deny the possibility of animal re-creation (and de-extinction) or give up, as resuscitation scientists propose, the view that extinction is forever.

The possibility of re-creating species

The idea of de-extinction seems to rest on an assumption that species are the kind of entities that can be re-created. The assumption is not self-evidently acceptable, since species are sometimes understood as individuals (Williams 1992, pp. 321–2; Hull 1994) and individuals have been argued to be the kind of entities that are beyond recreation (Gunn 1991, p. 301; Delord 2007, p. 660).

The view of species as individuals can be contrasted with a view of species as classes (Dupré 1992, p. 312; Levine 2001, pp. 326–7; Garvey 2007, p. 149; Elder 2008, p. 339). According to the latter, the more common view, individual organisms are members of a class called 'species'. Species are then seen as abstract categories with concrete instances (Powell 2011, p. 606). Some philosophers further hold that species are not classes of any kind, but natural kinds that exist in nature independent of their discovery or naming by humans (Elder 2008). Extinction can then be understood as a happening in which a class with members (that is a species) becomes a null class, in other words a class without members (Gunn 1991, 299). That a class with members becomes a null class and then later a class with members again is possible and common. A club or other kind of social institution consisting of members, for example, may occasionally become a null class and then later a class with members again. Thus, a view of species as a class, as such, is compatible with the idea of de-extinction. De-extinction is then understood as a happening in which a null class (that has had members before) turns back to a class with members (Powell 2011, p. 607).

According to the former view, 'species are individuals located in space and time with organisms as their constituent parts' (ibid., p. 606). Species are then understood as distributed individuals that are born, go through changes and die much like individual organisms do (Levine 2001, p. 327; Powell 2011, pp. 606–7). The view of species as individuals coheres with the idea of extinction as a non-desirable happening, a real loss, and the end and death of a species. This view, however, is not compatible with the idea of de-extinction if we accept the view that de-extinction is a form of re-creation and that individuals cannot be recreated (Gunn 1991, p. 301).

Gunn (ibid.) argues against the possibility of re-creating individuals by noting the identity of clones. Cloning an individual human being, Charles Darwin for example, does not imply a re-creation of Charles Darwin but the creation of a new human being who is genetically identical to Darwin. Even though it is possible to produce genetic copies of individuals, individuals themselves cannot be re-created. Thus, as long as species are understood as individuals, the possibility of re-creating them is excluded. It may then be possible to produce copies of species, but it is not possible to re-create the real and authentic original species (Delord 2007, p. 662; Powell 2011, p. 607). Birds born from de-extinction procedures may well be, according to this line of thought, genetically identical to passenger-pigeons and at the same time be parts of an individual species distinct from the lost passenger-pigeon. Thus, this view fits well with the first possible implication of the irreversibility of extinction and the idea that animals born from de-extinction procedures should be seen as copies or even as fakes.

It is easy to agree that the cloning of Charles Darwin or any other dead person is not equal to re-creating him or her. However, it is less certain that this implies the impossibility of de-extinction. After all, we usually have no problems in accepting that, even though a clone of Charles Darwin is not Charles Darwin, his clone, nevertheless, is a human being (Garvey 2007, p. 151). Analogously, Dolly the clone is commonly accepted as a sheep, not just a copy of sheep. As long as cloning is not seen as a threat to species identity in these cases, the opponents of the possibility of de-extinction need to explain why it is so in the context of de-extinction (Delord 2007, p. 662). If the only difference between Dolly and the clone of Celia, the last bucardo, lies in the existence of (other) members of the species cloned, the relevance of this difference to species identity needs to be conceptualized by the opponents of de-extinction. Brian Garvey (2007, p. 151) goes as far as to argue that the intuition that a cloned Tasmanian tiger is a real and authentic Tasmanian tiger implies that species 'do not behave exactly like the things that are undoubtedly individual'. Thus, according to him, the incompatibility of the view of species as individuals and the view that species can be re-created can lead into questioning the individuality of species rather than abandoning the possibility of re-creation. Species could then be seen either as classes or as pseudo-individuals – entities that resemble individuals to some degree but differ from them in some relevant way (Powell 2011, p. 608).

It is possible to argue for the compatibility of de-extinction and the view of species as individuals by claiming that de-extinction should not be seen as a form of re-creation but rather as an instance of resurrection or reviving. All three terms 'recreation', 'resurrection' and 'reviving' are quite commonly used in the context of de-extinction (see for example Delord 2007, p. 659; Redford et al. 2013, p. 3; Zimmer 2013, pp. 37, 41 for recreation; Delord 2007, p. 659; Silverman 2012; Stone 2013, p. 19; Zimmer 2013, p. 32 for resurrection; and Rosen 2012; Sherkow and Greely 2013, p. 32; Stone 2013, p. 19; Zimmer 2013, pp. 30, 33 for revival). These terms differ dramatically regarding their connotations. As noted above, the possibility of recreating an individual may be dubious, but there are no conceptual obstacles for accepting the possibility of resurrecting or reviving an individual. Had dying Charles Darwin been resurrected by a skilled medical group, the person brought back to life would certainly still have been Charles Darwin. Analogously, if a skilled group of scientists using some imaginary method could have revived Celia, the last bucardo who died in 2000 when a tree fell on her, they would have brought Celia back to life, not just any bucardo (or any animal). It seems fair to say, provided that species are understood as individuals, that they would also have brought another individual, the bucardo species, back to life. Now the crucial question is: can and should also the cases in which a new individual animal is brought to life by cloning, genetic engineering or back breeding be described as a case of resurrecting or reviving a species? If that is accepted, de-extinction can be seen as resurrection and reviving but not as re-creation. And since resurrection and reviving of individuals is conceptually possible, the view of species as individuals need not be incompatible with the idea of de-extinction.

To conclude, seeing species as a class is consistent with the idea of de-extinction. Moreover, the view of species as individuals does not necessarily contradict the idea of de-extinction, provided that species identity is never compromised in cloning or that, instead of re-creation, de-extinction can be understood as resurrection or reviving of the species.

Changes in animals and the loss of species identity

Genetic modification brings about, and is aimed to bring about, changes in its target. The idea of avoiding species loss by genetically modifying endangered animals so as to be more likely to survive in their changing environment rests on the assumption that some alterations – and some alterations following from the use of gene technologies – are minor enough not to compromise the identity of the animal so modified. The same assumption can be found behind all those de-extinction projects in which the production of animals which are not totally, but closely, identical to the originals is seen as a success and meeting the aim of the project.

Actually, the whole business of genetic modification seems to rest on a similar assumption. Genetically modified organisms are commonly seen as variants of the species in question: something with novel qualities but yet still authentic in the sense of belonging to the sphere of the original. Thus, it is not uncommon to discuss, as biologist Philip W. Hedrick (2001, p. 841) does, the possibility of transgenes spreading 'from GM organisms into natural populations of *the same* or related species' (emphasis added). At the same time, it is usually accepted that some ways of applying modern biotechnologies may compromise the species identity of their target (Savulescu 2011, pp. 656–7). Following these two commitments, a genetically modified Sumatran tiger that is more able to survive in its changing environment is still a Sumatran tiger, provided that the changes made to it are not too fundamental. Philosopher Keekok Lee rejects this way of thought by challenging the first assumption:

> On the surface, it appears that the [transgenic] animal is carrying out its own telos as its biological mechanisms remain intact; but if the implications of being a transgenic organism are fully teased out, the appearance of normality vanishes. This can be brought out by posing the question ... about its identity. There are two possible ways of

> answering the question 'what is it?'. One way is simply to say that it is still a cow, a tomato plant or whatever, which happens to produce a human protein in her milk or which happens to be able to withstand frost. The other is to say that it differs so fundamentally from a normal non-transgenic cow or tomato plant that it would be misleading to say *simpliciter* that it is a common or garden variety of cow or tomato plant. One could perhaps call it a Tgcow... or a Tgtomato plant. (Lee 2003a, p. 154; see also Lee 2003b, p. 8)

According to Lee (2003a, pp. 154–5), adopting the view of a transgenic cow or tomato plant as variants of cow and tomato is to go for appearances only and implies a failure to realize the inauthenticity resulting from the technological human intervention. Through genetic modification and presumably also through de-extinction technologies, an organism loses a feature integral to its identity – that is being born through evolution and natural processes and not being technologically produced. Following Lee's ideas, genetically modified Sumatran tigers are not real tigers, because of the very fact that they are genetically modified.

Analogously, birds born from the de-extinction procedures can never be real and authentic passenger-pigeons, even when genetically, morphologically and functionally identical to the original species. The history and origin of entities matter to their identity, and the technological human involvement in the history of these animals makes them distinct from the original animals. Thus, at best, they can only be human-made copies.

The view that human involvement matters to species (or subspecies) identity is not foreign to biology. It is hardly uncommon to give different scientific names to domestic animals and their wild counterparts (Price 2002, pp. 3–4). Even though scientists disagree as to how the distinction between domestic and wild animals should be expressed in nomenclature (ibid.; Gentry et al. 2004), they quite generally agree on the acceptability of expressing it. Moreover, human dependence is among the most commonly set criteria for being domestic. Domestic animals, as well as animals born from de-extinction and genetic modification, have their breeding, feeding and territory under human control (Isaac 1970, p. 20; Clutton-Brock 1989, p. 7; Gentry et al. 2004, p. 645). However, human dependence is usually not seen as sufficient for being domestic. Morphological changes resulting from the process of domestication (Isaac 1970, p. 20; Clutton-Brock 1989, p. 7; Jeffries 1997, p. 84; Price 2002, pp. 10–11; Gentry et al. 2004, p. 645) and sometimes also tameness (Isaac 1970, p. 20; Gentry et al. 2004, p. 645) are further requirements for an animal to be domestic. It is unlikely that all the outcomes of de-extinction and genetic modification will fulfil this set of criteria. Nevertheless, it may well be justified to consider some animals brought into existence in those ways as domestic ones. One such case might be genetically modified 'wild' pets (see Chapter 5): tigers, lions and other charismatic animals that have been genetically modified to be suitable companion animals in urban surroundings. According to the criteria presented above and provided that the morphological changes are deep enough, these animals are domestic, and thus there are reasons to see them as distinct from the original species and not authentic or real representatives of it.

That most animals born from genetic modification and de-extinction procedures fail to be domestic does not imply their being authentic members of the original species. There is no single generally accepted concept of 'species' (Queiroz 1998, p. 57; Powell 2011, p. 604), and thus there is no single generally accepted set of criteria for determining the species identity of animals born from genetic modification or de-extinction procedures. Some concepts of species, such as the so-called concept of biological species, according to which animals belong to the same species if they are actually or potentially interbreeding (Queiroz 1998, p. 58; Maclaurin and Sterelny 2008, p. 32; Powell 2011, p. 605), rely more on the actual properties of animals. Others resemble Lee's view in being more history-oriented, even to the point of stating that 'nothing that does not share the historical origin of the kind can be a member of the kind' (Griffiths 1999, p. 219). Species are then seen as a lineage of populations and organisms that is distinct from other lineages and has its own history and evolutionary tendencies (Queiroz 1998, p. 58; Maclaurin and Sterelny 2008, p. 33; Powell 2011, pp. 604, 606). According to this line of thought:

> no biologist would suppose that if organisms arose on some distant planet that were morphologically and physiologically just like house cats here on earth, and that those arose through natural selection from just the same sorts of ancestors as those from which our house cats arose, in response to just the same sort of selectional pressures involving just the same sorts of environing organisms and physical environments, and just the same sort of competition (for mates, for food, etc.) among the members of the species, those members would then be members of the species *Felis domesticus*... Biologists would judge that the bare numerical difference in origins entailed that the

> organisms on the distant planet were not members of the species *Felis Domesticus*. (Elder 2008, p. 359)

The crucial question arising from the history-oriented concept of species concerns the nature of the continuance and connection necessary for membership of the species. The link between original passenger-pigeons and the birds born from de-extinction procedures should be strong enough for the latter to belong to the same species as the former. Brian Garvey discusses this in length with respect to the imaginary scene of *Jurassic Park*:

> There is still some reason to say: yes, they [dinosaurs produced by de-extinction procedures] are members of the species *Tyrannosaurus rex*. The hypothetical scenario in Crichton's novels is that the new creatures were produced using DNA from actual, undoubted tyrannosaurus. So there is a causal connection between the undoubted tyrannosaurus and the new creatures, albeit a more tenuous causal connection than in the normal case... Suppose that, instead of using DNA from tyrannosaurus, the scientists synthesised DNA of the right kind using the raw material cytosine, guanine and so forth. Would we *now* say the new creatures were tyrannosaurus? Maybe; maybe not. On the one hand, the causal connection that was there on Crichton's scenario is no longer there. On the other, the scientists have *copied* tyrannosaur DNA, so there is still a causal connection, albeit one that is even more tenuous... And it is likely that in... the species case there are borderline scenarios where we would find it difficult to decide either way. (Garvey 2007, pp. 150–1)

It is easy to agree with Garvey's judgement that de-extinction and the genetic modification of endangered species often operate in the grey area in which decisions regarding sufficient continuance and connection are not easy. Despite and because of this, proponents of de-extinction, and especially scientists who claim themselves to work towards the goal of de-extinction, need to acknowledge these theoretical challenges. They can justify their work only if they can convince others that the outcomes of their work are what they claim them to be – real and authentic members of the formerly extinct species. They further need to realize that the concepts of species that are oriented towards non-historical properties of animals offer an easy way out only when one succeeds in producing animals perfectly identical to the original ones. In all other cases relying on them just leads into equally challenging questions about the degree of sufficient similarity regarding genetic, morphological and functional characteristics of the past and present animals.

The quality meaning of authenticity: the value of artificiality

One reason why Lee considers the origin of animals central to their identity is the artificiality and loss of naturalness following on from the involvement of human technology (Lee 2003a, pp. 1, 4–6). The distinction between natural and artificial has also played a central role in the ethical discussion concerning the so-called restored ecosystems (see for example Elliot 1982, 1997; Gunn 1991; Jordan 1993; Katz 1997, 2012; Lo 1999). A restored ecosystem is an outcome of 'the process of assisting the recovery of an ecosystem that has been degraded, damaged, or destroyed' (Welch and Cooke 1987). A restored ecosystem is similar to the ecosystem that was on the site before human beings caused disturbances; or is similar to the ecosystem inferred to have been on the site had those harms not taken place. The higher the degree of similarity, the more perfect the restoration (Angermeier and Karr 1994, p. 695; Higgs 1997, p. 343). As the terms 'degraded', 'damaged' and 'destroyed' in the above description hint, ecological restoration comes in degrees. At its minimum it may merely mean cleaning up anthropogenic trash and litter from an ecosystem. At the other end of the spectrum are cases in which an ecosystem is rebuilt after it has been destroyed by such human activities as mining (Katz 2012, p. 69). The latter kind of ecosystem restoration resembles de-extinction to a high degree. The aim in both cases is to bring a biological unit that has been lost as a result of destructive human activities back to existence. Thus, arguments concerning artificiality, authenticity and the value of restored ecosystems may be interesting and relevant to the question about the authenticity of animals born from de-extinction procedures. It has been argued that, because of their artificiality, restored ecosystems lack some value that the original authentic ecosystems had (Elliot 1982, 1997; Katz 1997, 2012). It might be suggested, analogously, that the animals brought into being by de-extinction lack some value and authenticity that the original members of the species had.

The question here concerns the quality meaning of authenticity. As noted at the beginning of this chapter, the quality meaning of authenticity is one of the two main meanings of authenticity and it concerns quality, typicality and the value of entities. This second sense of authenticity differs from the first one in leaving open the possibility that animals born from the procedures of de-extinction are members of the original species and at the same time less authentic than animals born through natural processes. The hypothesis then is that the artificial origin of the animals, the fact that they are produced by human beings, makes them less valuable and thus less authentic than the original members of their species (who were born independently of human beings).

From its very beginning the philosophical discussion on restored ecosystems has focused on pointing out that the history and origin of entities matter to their value. In his 'Faking Nature' Robert Elliot (1982, pp. 85–6) presents examples from the sphere of art. Even a perfect replica, one that the best experts can hardly tell apart from the original art work, lacks some value that the original art work has. Similarly, finding out horrible facts about the origin of an object of art, for example that it is made of the bone of a person killed especially for that purpose, changes one's evaluations about that object (see also Radford 1978, p. 76; Katz 2012, p. 68). The opponents of the restoration thesis – the view that the full value of a destroyed natural area can be returned by successful restoration – argue that ecosystems are analogical to works of art in this respect. Part of the value of nature's ecosystems comes from their being outcomes of natural evolution and independent of human agency. In other words, (part of) the value comes from their being authentic:

> Origin and historical continuity – what we might call *authenticity* – are crucial elements in the determination of the value of an art work... Shifting back to the restoration on natural systems, the elements of origin, historical continuity, and *authenticity* continue to play a decisive role in determination of value. (Katz 2012, p. 70, emphasis added)

> What people value in undeveloped nature is its natural history separate from human causation and activity. In an area that has been modified by human action there is a different causal history. Thus, even a perfect ecological restoration lacks the value of the original natural system it is re-creating, for the restoration was the product of human action. (Ibid., pp. 68–9)

Restored ecosystems and animals that have come into existence through de-extinction procedures are similar regarding their anthropogenic origin. Both have been brought into existence by human beings, and both are meant to be as similar as possible to a lost natural entity. Both are brought into existence by modern technologies, but neither of them can be said to be of human design. Rather they are human made copies of natural entities. Thus, it is fair to suppose that if restored ecosystems as a result of their artificiality lack some authenticity and value that the original undisturbed ecosystems have, then animals born from de-extinction also lack the value and authenticity in question. The analogy seems to concern, first and foremost, individuals of the first generation produced by the de-extinction procedures. When it comes to possible descendants of these animals, and possible populations formed by these animals, the analogy becomes weaker when the human involvement ceases and natural (that is human independent) processes take over. Nevertheless, even then, because the first generation animals were brought to existence by human technologies and natural processes then take over in these living artefacts, they are not equivalent to wild animal individuals or species (for similar views regarding restored ecosystems see ibid., p. 75).

It is more controversial as to whether the argument against the restoration thesis can be applied to the case of genetic modification too. Even though intentional human-made modifications are necessary for something to be an artefact (Elliot 1997, p. 123; Dilworth 2001, p. 345), mere modifications of an entity are not sufficient for turning it into an artefact (Siipi 2005, pp. 69–70). Something can be an artefact only if it has been intentionally produced and brought into existence by human beings (Hilpinen 1995, p. 138; Lee 2003a, p. 4). Thus, repairing a damaged (but not destroyed) ecosystem does not turn it into an artefact. Similarly, a minor modification of an animal (the kind of modification that does not lead it to lose its identity) is not sufficient for it to be an artefact. Only when human activities lead to the creation of something new that is distinct from its raw materials (that is animals of the modified species) is the outcome an artefact (Siipi 2005, p. 88). Thus, depending on how fundamental the changes made are, it may be possible to view genetically modified animals as variations and thus still as instances of the original.

Substitutes

Accepting the view that animals born from de-extinction or genetic modification are not authentic does not imply that these animals should not be brought into existence. Because of the intention to deceive imbedded in them, fakes are prima facie morally suspect, but other forms of inauthenticity are not morally problematic as such. They may be found acceptable or even desirable, especially in cases where the possibility of having authentic entities is excluded. In the context of de-extinction, and often also in the case of genetic modification of endangered species, the choice is not between authentic and inauthentic animals. Authenticity has already been lost or it is likely to be lost in the near future, and the choice is between having an inauthentic animal and not having that kind of animal at all. Should we then, in the name of favouring authenticity, rather choose not to have that kind of animal? The answer at least partly depends on which of the main meanings of authenticity is under discussion. With respect to the identity meaning, the question comes down to a choice between the following two alternatives: first, not having x (for example the passenger-pigeon) or anything of its kind; second, having y that resembles x (y is, for example, a bird that is not a passenger-pigeon but resembles one to some degree). It is not self-evident that having y should then be favoured. Since it is not x, bringing it to existence cannot merely be justified by the anthropogenic loss of x. Regarding the quality meaning of authenticity, the question concerns the choice between not having x and having a poor or lousy x. This may be an easier choice, and it might be claimed that as long as the existence of x is desirable, it is prima facie better to have a lousy x than no x at all.

However, even when we consider animals born from de-extinction and genetic modification as inauthentic in the identity meaning of the term, we may well ask whether they could serve as substitutes of the original authentic animals. The idea of substitutes is in an interesting way related to the ideas of the real and original. Substitutes are always substitutes of something, and being a substitute is then strongly dependent on the idea of something else being real and authentic. It is common to all substitutes that they are considered as less real (instances of something) than the authentic entity they substitute (Bergin 2009, p. 260). Substitutes share some (but not all) of their qualities with the entity they are substituting. Yet, they are in some sense less perfect and real than the authentic entity (Siipi 2013). Sometimes, but not always, substitutes differ from the real and authentic entities in being artificial. Most importantly, substitutes are at least to some degree capable of fulfilling the function or task of the authentic entity. If that kind of function or task can be found for the animals born from de-extinction and genetic modification, bringing them to existence might be justified by their role as substitutes. One such case might be the so-called Pleistocene rewilding – the idea of reintroducing large wild vertebrates into North America (and other areas, including Europe), from which they have died out. It has often been suggested that such a project would make use of closely related species (see for instance Donlan et al. 2005, p. 913; Svenning 2007). More ambitious projects that aim to a higher degree of similarity might involve animals born through de-extinction. Moreover, animals born from de-extinction, and preservation aimed at through genetic modification, might serve as useful substitutes in different kinds of scientific projects and biodiversity projects.

Conclusion

To conclude, the idea that de-extinction and the genetic modification of endangered animals produce authentic passenger-pigeons, Sumatran tigers and other animals rests on contestable views concerning species, identity and extinction. This, of course, does not imply that the animals produced are not authentic. The conceptions and views behind them are held by many, and they may turn out to be acceptable and correct. Nevertheless, contrasting views are also strongly supported, and thus the opponents as well as the proponents of de-extinction and genetic modification of endangered animals, especially scientists working on the matter, need to acknowledge, contest and justify the theoretical views behind in their policies and views.

The view that animals born from de-extinction procedures and genetic modification are not authentic members of the original species does not as such imply that these procedures should not be carried out. The animals born from them may, at least in some cases, serve as substitutes of the original. Alternatively they might be accepted as new valuable forms of biodiversity that contribute to ecosystem services and to attaining scientific knowledge. Despite the possible value and usefulness of inauthentic animals, the views concerning the authenticity of these animals are far from insignificant. When the view about the authenticity of animals produced by de-extinction and genetic modification is adequately justified, it offers strong support for applying these technologies. A well-justified view on the inauthenticity of these animals, on the other hand, may offer a way of solving a commonly presented concern that if species can be re-created then people will stop worrying about extinction:

> Current protection of endangered and threatened species owes much to the argument of irreversibility. If extinctions – particularly

> extinctions where tissue samples are readily available – are not forever, preservation of today's species may not seem as important. (Sherkow and Greely 2013, p. 33)
>
> Another worry is that bringing back any extinct species – pigeon, frog, or mammoth – could undercut efforts to protect habitats for endangered species. (Stone 2013, p. 19)
>
> Might we face the moral hazard whereby confidence in our ability to recreate extinct species undermines our willingness to conserve naturally occurring biodiversity? (Redford et al. 2013, p. 3)
>
> Cloning, such as transgenics, is a glamorous technology, and there is a danger of creating the false impression in the mind of a technology-infatuated public that it offers an easy, high-tech solution to the problem of extinction. (Ehrenfeld 2006, p. 731)

If the view about the inauthenticity of animals produced by procedures of de-extinction or genetic modification is accepted, appeals to inauthenticity (in both main senses of the term) may offer an argument against the scenarios presented. According to this line of thought, since the outcomes of de-extinction and genetic modification are not authentic, conservation and preservation aiming for extinction avoidance are always superior to these technological solutions and should be favoured over them. Thus, if the claim about inauthenticity is accepted, de-extinction and the genetic modification of endangered species cannot replace the current forms of conservation.

References

Angermeier, A.L. and J.R. Karr (1994) 'Biological Integrity versus Biological Diversity as Policy Directives: Protecting Biotic Resources' *BioScience*, 44, 690–7.

Bergin, L.A. (2009) 'Latina Feminist Metaphysics and Genetically Engineered Foods', *Journal of Agricultural and Environmental Ethics*, 22, 257–71.

Clutton-Brock, J. (1989) 'Introduction' in J. Clutton-Brock (ed.) *The Walking Ladder: Patterns of Domestication, Pastoralism, and Predation* (London: Unwin Hyman).

Delord, J. (2007) 'The Nature of Extinction', *Studies in History and Philosophy of Biological and Biomedical Sciences*, 38, 656–67.

Dilwoth, J. (2001) 'A Representational Theory of Artefacts and Artworks' *British Journal of Aesthetics*, 41, 353–70.

Donlan, J. et al. (2005) 'Re-wilding North America', *Nature*, 436, 913–14.

Dupré, J. (1992) 'Species: Theoretical Contexts' in F. Keller and E.A. Lloyd (eds) *Keywords in Evolutionary Biology* (Cambridge, MT: Harvard University Press).

Ehrenfeld, D. (2006) 'Transgenics and Vertebrate Cloning as Tools for Species Conservation', *Conservation Biology*, 20(3), 723–32.

Elder, C.L. (2008) 'Biological Species Are Natural Kinds', *The Southern Journal of Philosophy*, XLVI, 339–62.

Elliot, R. (1982) 'Faking Nature', *Inquiry*, 25, 81–93.

Elliot, R. (1997) *Faking Nature: The Ethics of Environmental Restoration* (London: Routledge).

Erler, A. (2011) 'Does Memory Modification Threaten Our Authenticity?', *Neuroethics*, 4(3), 235–49.

Garvey, B. (2007) *Philosophy of Biology* (Stocksfield: Acumen Publishing).

Gentry, A. et al. (2003) 'The Naming of Wild Animal Species and their Domestic Derivatives', *Journal of Archeological Science*, 31, 645–51.

Gentry, A. et al. (2004) 'The Naming of Wild Animal Species and their Domestic Derivatives', *Journal of Archaeological Science*, 31, 645–51.

Gilmore, J.H. and B.J. Pine II (2007) *Authenticity* (Boston: Harvard Business School Press).

Griffiths, P.E. (1999) 'Squaring the Circle: Natural Kinds with Historical Essences' in R.A. Wilson (ed.) *Species: New Interdisciplinary Essays* (Cambridge, MT: Bradford Book).

Gunn, A.S. (1991) 'The Restoration of Species and Natural Environments', *Environmental Ethics*, 13, 291–312.

Hedrick, P.W. (2001) 'Invasion of Transgenes from Salmon or Other Genetically Modified Organisms into Natural Populations', *Canadian Journal of Fisheries and Aquatic Sciences*, 58, 841–4.

Higgs, E.S. (1997) 'What is Good Ecological Restoration?', *Conservation Biology*, 11, 338–48.

Hilpinen, R. (1995) 'Belief Systems as Artifacts', *The Monist*, 78, 136–55.

Hull, D.L. (1994) 'A Matter of Individuality', in E. Sober (ed.) *Conceptual Issues in Evolutionary Biology* (Cambridge, MT: MIT Press).

Isaac, E. (1970) *Geography of Domestication* (Englewood Cliffs: Prentice Hall).

Jeffries, M.J. (1997) *Biodiversity and Conservation* (London: Routledge).

Jordan, W.R. (1993) 'The Ghosts in the Forest', *Restoration and Management Notes*, 11(1), 3–4.

Katz, E. (1997) *Nature as Subject: Human Obligation and Natural Community* (Lanham, MD: Rowman & Littlefield).

Katz, E. (2012) 'Further Adventures in the Case against Restoration', *Environmental Ethics*, 34, 67–97.

Lee, K. (2003a) *Philosophy and Revolutions in Genetics: Deep Science and Deep Technology* (Basingstoke: Palgrave Macmillan).

Lee, K. (2003b) 'Patenting and Transgenic Organisms: A Philosophical Exploration', *Techné: Journal of the Society for Philosophy and Technology*, 6(3), 1–16.

Levine, A. (2001) 'Individualism, Type Specimens, and the Scrutability of Species Membership', *Biology and Philosophy*, 16, 325–38.

Lo, Y.-S. (1999) 'Natural and Artifactual: Restored Nature as Subject', *Environmental Ethics*, 21, 247–66.

MacLaurin, J. and K. Sterelny (2008) *What is Biodiversity?* (Chicago: University of Chicago Press).

Powell, R. (2011) 'On the Nature of Species and the Moral Significance of their Extinction' in T.L. Beauchamp and R.G. Frey (eds) *The Oxford Handbook of Animal Ethics* (Oxford: Oxford University Press).
Price, E.O. (2002) *Animal Domestication and Behavior* (New York: CABI Publishing).
Queiroz, K. de (1998) 'The General Lineage Concept of Species, Species Criteria, and the Process of Speciation' in D.J. Howard and S.H. Berlocher (eds) *Endless Forms: Species and Speciation* (Oxford: Oxford University Press).
Radford, C. (1978) 'Fakes', *Mind*, 87(345), 66–76.
Redford, K.H. et al. (2013) 'Synthetic Biology and Conservation of Nature: Wicked Problems and Wicked Solutions', *PLOS Biology*, 11(4), 1–4.
Revive & Restore (2013) http://longnow.org/revive/ (accessed 27 August 2013).
Rosen, R.J. (2012) 'Assuming We Develop the Capability, Should We Bring Back Extinct Species?', *The Atlantic*, 7 August.
Savulescu, J. (2011) 'Genetically Modified Animals: Should There Be Limits to Engineering the Animal Kingdom's Extinction?' in T.L. Beauchamp and R.G. Frey (eds) *The Oxford Handbook of Animal Ethics* (Oxford: Oxford University Press).
Sherkow, J.S. and H.T. Greely (2013) 'What If Extinction Is Not Forever?', *Science*, 340, 32–3.
Siipi, H. (2005) *Naturalness, Unnaturalness, and Artifactuality in Bioethical Argumentation* (Turku: University of Turku).
Siipi, H. (2013) 'Is Natural Food Healthy?', *Journal of Agricultural and Environmental Ethics*, 26, 797–812.
Silverman, J. (2012) 'Resurrecting the Dodo and Other Extinct Creatures', http://science.howstuffworks.com/zoology/dodo1.htm (accessed 19 June 2013).
Stone, R. (2013) 'Fluttering From the Ashes', *Science*, 340, 19.
Svenning, J.-C. (2007) 'Slide Show: Bringing Back Europe's Prehistoric Beasts', *Scientific American*, 31 May.
Welch, E.B. and G.D. Cooke (1987) 'Lakes', in W.R. Jordan III et al. (eds) *Restoration Ecology: A Synthetic Approach to Ecological Research* (Cambridge: Cambridge University Press).
Williams, M. (1992) 'Species: Current Usages' in F. Keller and E.A. Lloyd (eds) *Keywords in Evolutionary Biology* (Cambridge, MT: Harvard University Press).
Zimmer, C. (2013) 'Bringing Them Back to Life', *National Geographic*, April, 28–41.

5

Bioengineered Domestication: 'Wild Pets' as Species Conservation?

Elisa Aaltola

Introduction

One of the notorious problems for species conservation is the rapid decline of natural habitats. This issue is faced also by species reintroduction programmes facilitated by zoos: even if one can keep a small population of animals alive, there may be no suitable place for them to be reintroduced to. One startling option is to accept the decline and adapt members of endangered species to live in domestic settings. As suggested by Kathy Rudy, conservation could take place via rendering wild animals into 'pets' – a project gathering momentum in the 'exotic animal trade'. In this chapter I will analyse this suggestion by exploring its possible implications from the viewpoint of animals, animality and species. First, what would the ensuing 'dewilding' mean: could one assert that an animal which has undergone significant behavioural alteration is still a member of the original species? Second, what consequences would the possibilities of genetic engineering, aimed at enhancing the sociability and welfare of the animals, have in this context: would the alterations in the genome not raise questions about the telos and ultimately species membership of the animals in question? Third, what implications would the animals' integrity and autonomy have: could we assert that the animal, or animality, was an object of respectful treatment within wild pet keeping? Fourth, how does the notion of bioengineered wild pet keeping relate to broader cultural meanings concerning essentialism and risk? It will be suggested that all of the above considerations force us to view wild pet keeping dressed as conservation in a critical light.

From zoos to homes

Although the greater majority of animals in zoos are not endangered, and although breeding programmes have faced severed difficulties, zoos often proclaim to take part in species conservation (Mullan and Marvin 1999; Snyder et al. 2002). This raises questions. Can wild animals kept in domestic settings be defined as representatives of wild species?

The question is twofold. First, the biology of these animals can be severely compromised. Gene pools face the risk of becoming impoverished, which again may lead to a situation where animals begin to lose many of those important features that render them successful in evolutionary terms. A related concern is that the behavior of the animals may undergo radical change, as those animals that can best cope in closely confined spaces are most likely to breed successfully, which again may mean that many of those behavioural features necessary in the wild disappear and so the animals become less viable in their natural surroundings (Marker and O'Brian 1989; Franklin and Frankham 1998; Snyder et al. 2002; Gautschi et al. 2003). In short, the animals' evolutionary development is radically altered, as animals adapt to human-made, extremely restricted environments, where the main adaptive pressures concern the capacity to tolerate both human presence and the lack of cognitive stimulation. Likewise, common problems with captive breeding include 'inbreeding depression, loss of genetic variation, and genetic adaptation to the captive environment' (Frankham 1995, p. 313; Snyder et al. 2002) and behavioural changes that compromise the animals' ability to survive in natural environments. Together with the tendency to select for tameness, these factors render future aspirations of reintroduction to the 'wild' often rather hopeless (Snyder 2002). Therefore, once wild animals enter human societies, the danger is that they cannot become the types of tigers, bears and elephants their ancestors used to be: the 'wildness' is lost for good.[1]

Second, as the animals lose their natural habitats, they risk becoming defined solely via their physical presence, rather than via their multi-layered relations with their respective surroundings. That is, species are no longer viewed as entities that exist in a particular dynamic with their environments, but rather become reduced to a collection of detached individuals. This feeds a notion of species as atomistic assortments of individual animals – a notion that is quite counterproductive from the viewpoint of further conservation efforts.

Of course, animal species are continuously adapting to anthropogenic pressures, and one could argue that the case of zoos is not qualitatively different from the type of change seen in particularly those animals who live close to human habitats – as is often suggested, the consequences of human actions are so widespread that almost all animal species on this planet have been impacted upon by them (Goudie 2013). Thus animals also become detached from their environments in the wild. Species are not rigid entities the identity of which stays the same throughout time – rather, they undergo continuous change, and therefore to ask for wild animal species to remain 'static' is misguided. Moreover, since only a metaphysic which separates human beings from other animals can provide the foundation for the concept of a pure human influence, then one cannot criticize species alteration or detachment from 'natural habitats' solely on account of their anthropogenic nature. Yet, the type of an impact a domesticated setting can have on the being or characteristics of wild animals can be drastic. It is the *level* of this impact, rather than human involvement per se, that forms the weightiest reason for doubting whether animals bred in zoos can stand as prototypes for their species.

There is one rather startling option. Perhaps wild species near extinction should be domesticated and embedded into human societies – that is, perhaps they should never be reintroduced back into the wild. 'Domestication' here would refer to rendering animals able to survive in anthropogenic environments, under human care, serving anthropocentric agendas.[2] Domestication would solve the aforementioned problems in one stroke: first, even drastic alterations in the animals' characteristics would no longer appear as negative, but rather necessary and even celebrated (allowing the animals to flourish in human societies); second, worries about loss of habitats and problems of reintroduction would no longer be relevant. In practice, what is being suggested is that wild species are rendered into domesticates or 'pets'.

Before the reader gulps in disbelief, it must be noted that there is growing support for 'dewilding' endangered wild species, as manifested in wild animal pet keeping (it is estimated that there are more big cats as pets in the USA (10–20,000 individuals) than there are tigers left in the wild).[3] Indeed, one trader in exotic wild pets commented in 2008 that 'the best way to save an animal's life – true conservation – is to create a market'.[4] This ethos finds limited support also amongst scholars. Thus, Kathy Rudy argues that domestication may be the only option in order to safeguard endangered species; 'given the shrinking space of the undeveloped "wild" world, those animals that can learn to live in connection with humans may have the best chance of survival' (Rudy 2011, p. 112). In fact, for her, keeping wild animals in the domestic setting is the only option that 'guarantees our children and their children will grow up on a planet that includes tigers, wolves, and other charismatic megafauna' (ibid., p. 148). Therefore, the argument is that there is ultimately nothing that can be done to prevent habitat destruction, and that the only solution is to domesticate endangered species into pets or (to use a more reflective term) companion animals.

According to Rudy, dewilding may spark moral awareness concerning the inherent value of wild animals, and thereby provoke respect or concern toward their flourishing. Her main argument is that *interaction* with other animals enables human beings to form moral, caring bonds with those animals. In other words, ethics stems from interaction and thus human beings will only truly be motivated to help wild species if they can see them face to face and form dynamic relations with them (on the normative ramifications of interaction, see also Smuts 1999). Hence, for Rudy, developing 'deep bonds' with wild animals will not only add richness to human existence, but will also help human beings to understand the moral value or significance of other species. She argues: 'my hope is that all animals, including wild ones, can become their own advocates through the personal experiences human share with them' (Rudy 2011, p. 113). In this way, dewilded animals would act as ambassadors for wild animals, and enable human beings to develop moral respect also toward those creatures still living on the planes of North Asia or the jungles of South America.

Rudy suggests quite simply that the solution is to render wild animals into 'pets', similar to dogs and cats. Although this option has a definite ring of insanity to it, Rudy is eager to emphasize its benefits in comparison to zoos. An often repeated criticism of zoos is that they do not offer a context in which one can relate to non-human animals as active agents instilled with various cognitive capacities and traits, who exist in relation to specific natural environments (Spotte 2006) – therefore, zoos quite simply offer a poor grasp of animality, and thereby poor grounds for normative awareness. Now, keeping wild animals as pets would definitely prohibit one from witnessing the animals in their native surroundings, but since at least ideally pet keeping involves interaction, it would seem that there would be hope of forming some understanding of animal agency. This forms one of the main justifications that Rudy uses to support her stance on wild pets: whereas zoos tend to offer us detached and passive animals, wild pet keeping can offer us intersubjective and active animals, with whom humans can have profound relationships and thereby equally profound learning experiences. In short, it is precisely pet keeping that enables the type of interaction that again can spark moral awareness and respect toward wild animals. Thus, if one accepts that interaction nourishes respectful attitudes toward other creatures (as Donna Haraway (2003) has shown in her *Companion Species Manifesto*, having a long history of interaction with dogs may have had a deep impact on the cognitive capacity of human beings to relate to and empathize with other animals), the argument could be made that being able to enter into dynamic relations with wild pets would encourage concern for not only the individual animals but also their respective species.[5]

In order for them to flourish as pets, wild animals would have to undergo substantial change. Their desire to roam free, to hunt, to form social groups with other animals, and (quite importantly) their desire to kill and possibly eat their human guardians, would have to be curtailed and ultimately eradicated. Moreover, in order to have a reasonable level of welfare, the animals would have to be altered so as to become much more modest when it comes to their own needs and interests: freedom to follow species-specific traits would, in most instances, quite simply become impossible to secure, which would require drastic changes in the physical and mental constitution of the animals. Therefore, the type of change Rudy is after is significant – it concerns altering some of the most integral aspects of animal nature. This is not an easy project to carry out, and would, in fact, require extremely careful breeding practices. Added pressure comes from the fact that this alteration would have to be quick: habitats are disappearing fast and traditional forms of breeding may simply be far too slow. The obvious option here would be to use bioengineering to speed up the process.

Via 'bioengineered domestication', wild animals could be rendered more able to participate in human societies. Both cloning and genetic modification emerge as options. Via cloning, wild animals could be 'copied' endlessly, even if reproduction in the domestic setting proved difficult – more importantly, those animals with advantageous features could be cloned. And with genetic modification, the very genomes of animals could be altered so that more desirable qualities would become prevalent, and the unwanted qualities 'knocked out'. It has to be noted that, for either of these methods to become viable, much more scientific development would have to take place. Such efforts of development are underway within animal agriculture, where bioengineering is explored as a method of optimizing animal production (Thompson 2007). Furthermore, cloning companion animals is becoming increasingly commonplace, particularly in the USA, and many transgenetic experimentations on pets are under way (including the fluorescent dog created at Seoul National University in 2009). To carry on the list, genetic modification of animals has increased exponentially within animal experimentation industries. Therefore, it does seem regrettably likely that the possibilities of bioengineering 'wild pets' will be investigated in the future.

The question becomes: is it justifiable to clone or genetically modify wild animals into 'pets' in order to keep their species alive? Using cloning and genetic modification as a part of species conservation has been suggested by many (see for instance Ryder and Benirschke 1997) – however, would 'bioengineered domestication' be justified?

Dewilding

The obvious criticism, mentioned above in the context of zoos, is that dewilded animals are not representatives of their original species, precisely because they have gone through such a fundamental process of alteration. The animals would be made, to phrase it simply, 'dumber', as they would lose the type of agency and intentionality required in wild environments, but which would be burdensome in the domestic setting. The animals would also be rendered significantly 'duller', as they would lose, for instance, their desire to roam and hunt. In order to dig deeper, 'domestication' should be kept separate from 'taming', the more usual method of approach toward near extinct wild species such as lions or wolves living in zoos. What Rudy's argument implies is that it is, indeed, domestication (of lineages) rather than taming (of individuals) that is required. When it comes to the effects of domestication, one common denominator is that the animals in question adapt to human influence: 'critically, all domesticates manifest a remarkable tolerance of proximity to (or outright lack of fear of) people' (Driscoll et al. 2009, p. 9972). It is noteworthy also that other 'anthropomorphic changes' are often momentous. In particular, it is common that domesticated animals have clear 'manifestations of neoteny' (ibid.), which means that they possess many juvenile characteristics, both physically and behaviourally. The obvious implication or risk here is that wild pet keeping reduces wild animals into juvenile cuties – a process similar to that of turning wolves into dogs. Therefore, according to the criticism, the end result would be caricatures of lions or gorillas, not 'lions' and 'gorillas' per se. This would also downplay the suggestion that dewilded animals would invite normative concern for their wild conspecifics: equally as interaction with dogs arguably teaches little respect toward wolves,[6] interaction with dewilded lions could fail to spark respect toward wild lions.

One way to approach the issue is to concentrate on 'wildness': bioengineered domesticates could not represent wild species, simply because they are not wild. However, there are problems with this claim. The line between 'wild' and 'domesticated' animals is blurry: as already argued, species are constantly undergoing change, and it is difficult to pinpoint the moment when a wild species has become domesticated (Driscoll et al. 2009). This is something that Rudy emphasizes, as she claims that the differentiation between 'wild' and 'domestic' is untenable. According to her, these categories are culturally produced rather than purely descriptive hard facts, and in the current context of rapid species extinction not the prudent way forward (Rudy 2011).

Another, perhaps more fruitful option is to concentrate on the level or intensity of the change itself: how fundamental or significant is the alteration that is being proposed? It would seem that, although drawing specific lines is difficult, we can easily pinpoint cases where change has been so significant that it warrants one to assert that a category shift has been made. The type of humanized change enabled by biotechnological intervention would seem potentially to belong to this class. We can assert that a tiger, which is less fearful of human beings, may still be a 'tiger', but that a tiger, which yearns for human company and lacks the desire to roam or hunt, is not. That is, the less an animal exhibits traits that are considered intrinsic to its species, the less reason there is to call it a member of that species. The line is fuzzy and relative, yet meaningful. Here it ought to be pointed out that defining 'species' even as a purely biological concept is notoriously difficult, and little agreement over the matter exists (Dupre 1999). One solution is to allow folk taxonomies more significance (Dupre 2002). This would be a combination of both biology (including emphasis on genomic similarities and viability) and folk taxonomies (including emphasis on behavioural similarities) that acts as the basis for this quantitative stance on species-category shifts.

To see just how extensive a change has been suggested, it is worthwhile paying closer scrutiny to Rudy's approach. Rudy supports a contractual interpretation of the relations between humans and the representatives of wild animal species. In her opinion, wild animals would form 'contracts' with human beings, and thus become proper members of human society. She talks of wild animals 'letting us tame them', and even quotes a poem where an animal is looking forward to the 'wonderful' life she will have after being tamed by human beings (rather horrifically, the poem ends: 'Please... Tame me!') (Rudy 2011, p. 151). Therefore, Rudy repeats the 'civilization claim' – according to which animals choose to be domesticated, and upon their decision enter under the legislation of human society, where they have to give away certain rights (most obviously, the right to freedom) and also accept certain duties (most obviously, the duty to bind by human rules). In effect, what Rudy is suggesting is the total transformation of wild animals into socialized creatures, the subjects of contracts. In effect, what is being proposed is not only the eradication of wildness, but also the nullification of the animals' independent agency. Animals would only exist in relation to, and dependent on, human beings. Here, animals would not only lose some of those qualities that make life in captivity difficult, but they would also be expected to become creatures capable of following anthropogenic rules. This implies a momentous transformation, which surely is significant enough to warrant claims of a category shift.

Representatives of species?

To have a deeper look at whether bioengineered wild pets can represent their original wild species, it is beneficial to explore the implications of bioengineering itself. One crucial point of call here is the term 'natural'. Are animals being altered in an 'unnatural' way, which again would make them poor models for their species? This is, of course, an age-old question within genethics (Siipi 2008). 'The argument from naturalness' maintains that it is wrong to interfere with the inherent nature or essence of a being. Often, the argument rests on teleology, as it is presumed that each animal has its own telos, which derives from a long evolutionary history and defines the essence of the animal, the fulfillment of which is necessary for its flourishing. Following this teleological logic, bioengineering would not only compromise the essence but also the welfare of the animals in question. Furthermore, it would be 'unnatural' because it would simultaneously alter their very telos to a degree that excludes them from their original species.

However, as often happens with arguments that rely on the term 'natural', problems are on the horizon. One common criticism is that neither genes nor animals have any 'essence'. That is, there is no inherent, core nature, which every animal of a given species follows (Thompson 2007). Not only species, but also genes, undergo constant change, particularly in the way they relate to one another, and thus the type of stability required for one to have an 'essence' is nowhere to be found. As has been argued by Bernard Rollin (1995) and Mark Sagoff (2001), the human act of altering the genes of non-human animals has already taken place, albeit in a more indirect form, for the whole duration of domestication. Unless one wants to declare all influence on other animals as 'unnatural', there seems to be little reason to reject genetic modification on these terms (see also Verhoog 2003). Yet, the argument from naturalness can also be defended. A lack of static essence does not necessarily mean a lack of telos. We can argue that given types of animals typically strive toward given modes of being, even if we accept that much variation and change happens within these modes. That is, telos can refer to a broad tendency which different individuals may exhibit in different ways and which is subject to alteration as time goes by. To put it simply: there is no reason to assume that telos is fixed in time or structurally uniform. What becomes relevant is how significant a change takes place – as argued above, considerable alteration would lay the potential grounds for asserting that one can no longer speak of the same species.[7]

However, perhaps there are instances where drastic modifications of the telos are justified, even if they lead to species fissure. Whereas above, emphasis was placed on essence, here the other factor within teleological concerns – flourishing – emerges as relevant. Let's look at a parallel case that concerns production animals. It has been argued that genetic modification of production animals, which aims at curtailing the animals' capacity to feel pain or otherwise feel ill, is justifiable (see Rollin 1995; Thompson 2007). In fact, Bernard Rollin has argued that since it is unlikely that methods of production will become more animal friendly (if anything, they are likely to get significantly worse), the moral thing to do is to map out ways in which we can reduce the animals' ability to suffer by changing their very biology. Now, this case is comparable with the case concerning genetic modification of wild animals. In the context of both, it is suggested that GM methodologies are justified, if they are necessary for the betterment of animal welfare.[8] For Rollin, what matters is that telos alteration must serve the interests of the animals. He puts forward the 'principle of conservation of welfare', which asserts that 'genetically engineered animals should be no worse off than the parent stock would be if they were not so engineered, and ideally should be better off' (Rollin 2006, p. 77). If applied to wild animals, this principle would suggest that they may be genetically altered, if doing so does not bring the animals any welfare disadvantage, and instead actively serves their welfare interests. That is, if a lion is happier as a docile, dependent and cognitively less able creature, genetic modification may not pose a problem. If indeed the only place where a lion can potentially flourish in this era of habitat destruction is human society then modifying it in ways that will render its existence within that society easier will surely be justified.

However, not everyone agrees with interfering with the biology of animals, even if doing so serves their welfare. Traci Warkentin has, in fact, argued that this option is nightmarish due to its wider implications. The most important of these is that animals would be, in practice, rendered into the type of 'mechanomorphic' creatures that anthropocentric illusions have for long argued animals to be (within mechanomorphia, animals are wrongly presumed to be biological mechanisms rather than cognitive individuals – see Crist 1999). Therefore, mechanomorphic misapprehensions would be made real: one would literally transform animals into non-cognitive creatures (Warkentin 2009). Paul S. Thompson has offered a similar argument. He uses Aldous Huxley's novel *Brave New World* as an example of a dystopia where human beings chemically alter their own minds so as to avoid being aware of the horrid state of the world. Even though the end result is less suffering, it would seem evident that the prize for avoiding suffering is overly high: cognitive ability and awareness of the surrounding world are lost. Thus, as Thompson points out, perhaps the only relevant question is not how we feel, but also what types of creatures we are – in other words, perhaps the integrity of the essence or minds of the animals triumphs over their welfare (Thompson 2007).

From this viewpoint, using biotechnological means to render wild animals more able to cope in human society would face the risk of mechanomorphizing animals. This risk has potentially disastrous moral implications. If, indeed, they were rendered cognitively less able, pet tigers and wolves would start to resemble the Cartesian automata that animals have historically been likened to. Since this mechanomorphic image has played a significant part in undermining the moral status of non-human animals (Crist 1999), the project of biotechnological domestication would lead to unwanted consequences: perhaps human beings would start to value wild animals less, not more. That is, whereas Rudy hopes that wild pets would spark respect toward animality, the very opposite could, in fact, happen. The loss of moral respect toward individual animals is easily accompanied by a loss of respect toward their species, as it could lay the ground for doubt as to why preserving collectives of cognitively unable animals is, in the end, so very important. That is, if Rudy's premise relies on moral psychology, the danger is that, on the level of such psychology, people would begin to see less value in other species.

It is worth emphasizing that, with the diminishment of their often brilliant cognition, wild animals would lose part of their ability to make sense of the world – that is, they would be stripped of their very ability to relate to reality in a clear, meaningful manner. This in itself is problematic, particularly from a teleological, virtue ethics, point of view. In her 'capability approach', Martha Nussbaum has argued that one foundational element in moral consideration toward others is supporting the fulfilment of their cognition. In other words, we are to help others to achieve their cognitive telos, to become 'who they are' as minded creatures. Furthermore, tampering with this telos amounts to a moral crime, for in so doing one takes away the 'selfhood' of the other creature. This applies fully in the context of non-human animals. Indeed, Nussbaum asserts that the way in which domestication of farmed animals has negatively affected their mental ability is, in itself, a moral wrong that should be corrected (Nussbaum 2004). Therefore, the claim is that safeguarding the capacities of animals is integrally related to respect toward the telos of animals and ultimately their 'selfhood', and any act of tampering is thereby immoral.[9]

It is worth noting one further risk: discrimination against those that are not genetically modified (Warren 2002). If indeed wild animals were 'civilized' and rendered more willing to enter into human–animal interactions, a danger is that their 'wild' counterparts, who are still hostile toward human beings, would be seen as inferior and even faulty in comparison. That is, if a wild lion refuses to form friendships with human beings, would it, in the dystopias of wild pets, not be viewed as imperfect, and even incapable? The worrying possibility is that the features of wild species would become unwanted and negative, and only humanized features would be deemed acceptable. Thereby, the very 'animality' of these species would be eradicated, and ultimately human beings would be taking part in an anthropocentric and anthropomorphic fantasy, in which all that we can see around us are human or human-friendly features. No diversity, no hostility, no awe in the face of difference. This, surely, would be the final blow to the survival of wild species, all spectacular in their specificity and difference.

Therefore, not only does the extent of the suggested bioengineered alteration pose problems, but also its further implications render it dubious to claim that wild pets could act as representatives of their species. The danger is that wild pets would invite mechanomorphia, moral detachment and loss of animal identity. From this viewpoint, wild pets would be, not ambassadors of their species, but rather 'bad copies'.

The normative relevance of creating such copies is evident. Whereas Rudy suggests that wild pets could invite respect toward animality, it would seem that, not only would their existence hinder such respect, but the very act of producing these pets could, in itself, be a sign of moral disregard.

Integrity, imagery and control

Warkentin proposes that the *integrity* of animals demands that they should not be genetically modified (Warkentin 2009; see also Kaiser 2005). Following a Kantian logic, one reason behind this concern appears to be loss of 'autonomy', a concept often linked to the integrity of individual beings. From this perspective, the integrity of the animals would be violated, because biotechnological domestication would tarnish the animals' capacity to be independent agents, able to direct their own actions according to their own intentionality. That is, the animals' capacity to be autonomous is at stake. As Henk Verhoog points out, the danger is that 'no respect is shown for [the animal's] independence or self-regulation, its "otherness"' (Verhoog 2003, p. 295). This problem is evident in the context of wild pets, whose being would be altered precisely so as to extinguish the more undesirable aspects of their intentionality and to replace independence with dependency.

Concern for the integrity and autonomy of animals is a familiar theme from the zoocentric approaches often evident in animal ethics. Yet, this concern does not need to revolve around individual animals, but can rather be expanded toward the level of species. Evidently one aspect of species survival is that individual members of the species are able to fend for themselves, and to fight for their own survival. Hence, autonomy – understood in a broad sense of the term – is a necessary requirement for the flourishing of species. Since the practice of keeping wild pets explicitly seeks to eradicate or at least lessen animal autonomy and replace it with dependency on human interaction, its ability to accommodate concern for species – and particularly species survival – becomes highly dubious. Here, it is not only the fact that the telos of animals is tampered with that causes concern, but also the fact that the very ability for survival in the wild is eradicated. One specific example of loss of autonomy concerns reproduction. Sarah Franklin argues that in particular cloned animals lose a core ingredient of autonomy: the capacity to reproduce independently (that is, the act of cloning sidelines independent reproduction) (Franklin 2007). A similar argument could be made in relation to bioengineered wild pets, which are no longer the products of the choices and intentionality of animal agency, but rather the choices and intentionality of the human scientist. Here, animals are prevented from having an impact on their own survival, via selecting the types of mates and places of birth that most probably ensure viable offspring. Therefore, perhaps the existence of wild species is rendered so integrally dependent on humanity (from birth to death and everything in between), and the autonomy previously included in them so forcefully curtailed, that it no longer makes sense to speak of 'species survival'. That is, perhaps with the almost complete loss of autonomy, comes the death of the species. (Following a Rolstonian environmental ethic with its emphasis on the integrity of species, the broader question becomes: can we talk of moral regard or respect for species, if its mode of existence is rendered dependent on human manipulation and support?)

Next to the issue of integrity. It is imperative that attention also be placed on the way in which bioengineering bears an impact on how animals are conceptualized. More specifically, what type of an *animal imagery* do these practices produce? The most obvious danger is the *commodification* of animals. If we follow Locke's well-known interpretation of property, human beings can, via the act of manipulation, declare species to be their own 'property'. To follow this logic, the more integral and obvious the manipulation, the more tightly the animal species becomes a human possession. Now, rendering species and individual animals into property is, in itself, highly problematic,[10] and these problems are made even weightier when we keep in mind that property can always be sold – that is, it can always be reduced into a commodity. This is not only troublesome from the viewpoint of animal ethics, but more broadly from the viewpoint of species.

As Mary Ann Warren argues, animals produced via genetic modification are easily seen as 'products' of human development, which further erodes moral concern for non-human animals (Warren 2002) and, arguably, their species. Thus, animals risk becoming not just sources of different commodities (such as meat, dairy or environmental experiences), but also commodities *in themselves*.[11] Franklin offers a similar view and maintains that bioengineered animals risk being patented and becoming trademarks owned by scientific corporations such as Monsanto, which again would enable extreme commodification, as animals become 'brands' that are sold like cars or televisions. What is central is no longer the individual, but the *prototype* that the individual represents (Franklin 2007). When it comes to wild pets, these are real risks. The very act of manipulating the genome of animals could be interpreted as a manifestation of one's property rights.

So if bioengineered wild animals came to be seen as the prototypes of their species, the scientist could claim to own the entire species: 'lions' would become property. Bioengineered lions and wolves could become brands owned by bio-corporations, patented commodities that the affluent flock to buy.[12] This state of affairs would be rather contradictory to species conservation, as the very mechanisms that have led to species extinction (consumption and ownership of nature) would control what is left of wild animals. Our very understanding of what wild species are would undergo a dramatic change. Independent beings would be replaced by dependent commodities, property that can be bought. Whether the commodified pets could ever invite respect and moral regard toward wild species is therefore seriously questionable.

This leads us to a further important issue, that of control. DNA represents a significant leap in the biological sciences, as it enables intervention and ultimately control over that which scientists used to observe merely from a distance (Turner 2002). The theme of control is also important in the context of wild species, as the cynical view is that the biotechnological methods of species conservation are forms of extreme human control, and thereby examples of anthropocentrism gone wild. Here, humans become the manipulators of wild species and ultimately hold excessive control (to use Aristotle's terms) in both *causal* and *formal* fashion, as both the origins and the characteristics of the animals are dictated by human beings. Whether such a framework of control could ever facilitate moral respect toward species seems very unlikely indeed.

Clones, codes and consequences

Finally, for the sake of the argument, let us presume that genetic modification would not be so extreme as to render talk of previously wild pets belonging to their original species to be invalid. Could pet wolves or bears represent their species, if their genome was relatively unaltered?

The cultural discourse concerning genes often implies a sense of essentialism and universality: DNA is understood to capture the universal essence of species or the very identity of individual beings (Warren 2002). In this vein, Katherine Hayles has maintained that part of the 'post-human' condition is the belief that information exists in itself, independently of material elements, and ultimately dictates all existence. That is, information is the secret key or code to the universe (Hayles 1999). According to Turner, it is particularly this approach that has opened the door for the commodification of biological information:

since science is so closely intertwined with corporative interests, scientific innovations are easily rendered into sellable commodities or products. Here, cloning becomes particularly relevant, as it rests on the (paradoxically unscientific) idea that there is, indeed, a universal code or independent information that can be transferred from one material body to the next, and that clones can be used to reap financial profit. That is, species become algorithms independent of bodily animals, and these algorithms can again be used as sellable products (Turner 2002).

However, as Warren points out, the idea of essentialism and universality is highly dubious. It is not only genes that make species or individuals into what they are but, rather, also the complex and intricate relations that they share with each other and the external world. That is, tigers and pandas are not only collections of genes, but foremost beings who have been shaped into what they are by their relations to other animals and their wider surroundings (Warren 2002). In this ethos, Holmes Rolston argues that 'a species is what is inseparable from its environment... It is not preservation of species but species in the system that we desire... Ex situ preservation, while it may save resources and souvenirs, it does not preserve the generative process intact' (Rolston 1993, p. 153). Rollin warns us of the type of reductionism that easily takes place here. There are many dimensions to explaining what type of beings animals are, and reference to their DNA is just one limited view which, when overly emphasized, may block out understanding of other, equally important explanatory frameworks (Rollin 2006). Therefore, there is no 'universal information', no 'immaterial code', that species could be reduced to. By simply safeguarding genes, species will not be saved. Simply to keep some representatives of species alive in artificial settings, as some type of 'gene vehicles', is by no means enough to ensure that the species itself is being kept alive. When you take the environment out of the equation, you are left with very little. Therefore, the type of commodities that wild species could become may have very little to do with actual 'species'. Sustaining a population of previously wild animals as pets is not species conservation, even if their genes were relatively intact.

Stephanie Turner has offered an interesting reading of the broader cultural and philosophical ramifications of cloning by using the *Jurassic Park* books as her reference point. She shows how, despite the 'ideology of the code', that is, the culturally produced belief that DNA is the secret algorithm to the reality, the very prominent fear is that the clones produced with this algorithm turn out to be the aforementioned 'bad copies' (Turner 2002). This fear is both practical and theoretical. Practical because the most notorious problem with cloning is that it does, indeed, tend to give birth to beings who are either badly disfigured or who – when seemingly perfect and healthy – suddenly die for no apparent reason. Theoretical because the fear is that there will be something sinister or twisted about the clones, that they will not behave as they should do, and that they will, like the dinosaurs of *Jurassic Park*, ultimately turn against their makers. This issue is also relevant in the context of wild pets.[13]

When separated from their environments, and when bioengineered so as to accommodate human societies, wild pets may turn out to be very different types of creatures than originally hoped. These 'bad copies' could have a devastating impact on a very practical level. They could begin to exhibit undesirable qualities, if the type of features that were manipulated into them had unpredictable consequences – a very likely event given the largely unknown, incredibly complex interrelatedness between genes. For example, genes associated with sociability could unexpectedly give birth to new diseases or render the animals incapable of reproduction in years to come. Therefore, on the individual level, wild pets could begin to portray unforeseen traits that had serious implications for the viability of the animals themselves, and for the health also of other animals and even human beings. Secondly, there could be hazardous environmental consequences. One often repeated argument highlights the possibility that modified animals may, in the end, play havoc on biodiversity. The very real threat is that the genetically modified animals, who were meant to preserve biodiversity and species, end up (rather paradoxically) having a destructive influence on those very entities. Here, species conservation would transgress into species destruction.

Hence, it can be argued that not only do bioengineered wild pets fail to act as tools for species preservation, but that they also may have harmful implications from the viewpoint of animal health and species and environmental protection. Belief in genes and codes risks blinding us to the practice of how species form and what the consequences of conserving species via bioengineering could be.

Conclusion

Bioengineered domestication offers one, albeit sci-fi like, dystopia of species preservation. Here, instead of demanding that human beings become more 'nature-orientated' in their modes of existence, and thereby more respectful toward other species and the wider ecological ramifications, the demand is that wild species need to become more 'human-orientated' and alter their modes of existence so as to fit into the demands of human societies. Yet, bioengineered domestication meets various difficulties. First, the significant extent to which it alters animals is problematic from the viewpoint of species preservation; second, it easily repeats the mechanomorphic understandings of animals that go against a broader moral concern over species extinction. Third, it violates the capacities of animals in a way that can be deemed morally reprehensible. Fourth, such domestication goes against the autonomy and, more broadly, the integrity of animals, and yields to a commodified understanding of wild species. Finally, it can lead to unwanted, destructive consequences for animality and species conservation. All these problems are significant, not only from the viewpoint of individual animals, but also from the viewpoint of species. The type of respect that the supporters of wild pets argue for cannot be attained as long as these problems stand.

But what could respect be grounded on? Warkentin emphasizes humility and wonder. We must become aware of how unpredictable and uncontrollable 'nature' (including non-human animals) is, and how it cannot be reduced to fit single codes or be viewed as a controllable commodity. That is, the autonomy of nature needs to be brought to the fore – an autonomy before which human beings are often powerless and secondary. From the perspective of species conservation, this would require that wild animals are kept as potent, able agents, relatively independent of human society. Their habitats may be rapidly vanishing, but the fruitful option is not to give up and demand that animals become human-orientated. Rather, a genuine push for human beings to become more animal-orientated is required, with a significantly more serious and strenuous emphasis placed on preserving their habitats than is currently being manifested.

Notes

1. Another fundamental problem is that wild habitats are quickly vanishing – in other words, soon there may be no place to relocate the animals to.
2. There is, of course, variance within 'domestication', with many semi-domesticated animals fulfilling only one or two of these three criteria (hence, reindeer do not live in human-built environments and rats are not under human care, nor do they serve human motives).
3. See National Geographic 2012, http://channel.nationalgeographic.com/wild/animal-intervention/articles/wild-at-home-exotic-animals-as-pets/ and Captive Wild Animal Protection Coalition, www.cwapc.org/education/download/BigCatFacts.pdf (accessed 1 August 2013).

4. See *The Collegian* 16 September 2008, www.kstatecollegian.com/2008/09/16/manhattan-native-promotes-conservation-through-exotic-pet-store/ (accessed 1 August 2013).
5. What about the wellbeing of the individual animals? Rudy emphasizes two issues: the welfare and safety of animals. For her, both are under danger in natural environments, whereas in the domestic setting both can be respected. In fact, Rudy maintains that wild animals can fare well as pets, and even have 'happy lives' as domesticated creatures. She acknowledges that they may get 'a little bored' (Rudy 2011, p. 130), but sidelines this issue on the premise that dogs, too, can get bored. For Rudy, 'love' is a main theme, and she argues that wild pets, too, can enter into loving relations with their human guardians, with the result that both the animals and humans in question gain more depth to their existence (Rudy brings forward a worry that animal rights discourse may lead into a world without bonds with other animals, and sees wild pet keeping as one solution). In fact, in Rudy's opinion, animals may gain much from human contact. Of course, this view is optimistic at best and absurd at worst, if one takes into account the fact that the greater majority of the specific tendencies and interests of wild animals' species would be frustrated. Indeed, Rudy describes a pet tiger that she visited and maintains that he was quite happy for the chance of being able to run around the perimeter of a yard 45 times a day – a conclusion that more sceptical readers will find quite worrying. Therefore, from the welfare viewpoint, Rudy's argument is highly incongruous and untenable.
6. Indeed, many 'nations of dog lovers' are also nations of mass hysteria and terror toward wolves (Finland being one example).
7. It is quite possible to state that, even if traditional breeding has also altered species, genetic alteration is none-the-less morally problematic. Here, the presupposition is that it is logically fallacious to follow the *tu quoque* form of reasoning, within which one type of an evil is justified by the existence of another.
8. It should be noted that later Rollin has changed his views, and argues that the eradication of positive cognitive features (that enhance welfare) is morally problematic (Rollin 2003).
9. Of course these capacities may change in the 'natural' setting too, but this is no justification for purposeful alteration (a parallel case would be suffering, which is frequent in the wild, but which human beings nonetheless have a duty to avoid causing).
10. See Gary Francione (2004) and his criticism of animals as property.
11. According to Rollin, commodification of animals does not necessarily pose a problem, for we can imagine farmers that would have cloned stock, but who would still treat their animals as individuals.
12. Alongside ecological factors, economic factors are relevant. As Warren points out, genetic modification is not just a scientific or moral issue, but is also deeply related to economic and political considerations. For instance, it may be only the economic and societal elite that can gain access to the more profound biotechnological innovations (Warren 2002). Now, this is extremely relevant in the context of wild pets, for surely the danger is that only the very rich will ever be able to keep tigers, elephants or bears as pets. If this is the case, are not nearly all the benefits suggested by Rudy only applicable

to 'the elite' rather than to society as a whole – that is, would it not just be the elite that gets to interact with animals, whereas those with less money would be utterly detached from these last remnants of wild species?

13. As Turner points out, Baudrillard has talked of the 'mania of origins', which refers to the desire to restore out of fear the reality that will otherwise be lost. Restoration often gains the form of rewriting or 'face-lifting', as the aim is to render reality perfect and stable (Baudrillard 1994). Yet, at the same time, the fear is that, underneath this perfection, something uglier emerges.

References

Baudrillard, J. (1994) *The Illusion of the End* (Stanford: Stanford University Press).

Crist, E. (1999) *Images of Animals: Anthropocentrism and Animal Mind* (Philadelphia: Temple University Press).

Driscoll, C., D. Macdonald and J. O'Brien (2009) 'From Wild Animals to Domestic Pets: An Evolutionary View of Domestication', *Proceedings of the National Academy of Sciences*, 106(1), 9971–8.

Dupre, J. (1999) 'On the Impossibility of a Monistic Account of Species' in Robert Wilson (ed.) *Species: Interdisciplinary Essays* (Cambridge, MA: MIT Press).

Dupre, J. (2002) *Humans and Other Animals* (Oxford: Clarendon Press).

Francione, G. (2004) 'Animals – Property or Persons?'. in C. Sunstein and M. Nussbaum (eds) *Animal Rights: Current Debates and New Directions* (Oxford: Oxford University Press).

Frankham, R. (1995) 'Conservation Genetics', *Annual Review of Genetics*, 29, 305–27.

Franklin, S. (2007) 'Dolly's Body: Gender, Genetics, and the New Genetic Capital' in L. Kalof and A. Fitzgerald (eds) *The Animals Reader: The Essential Classic and Contemporary Writings* (Oxford and New York: Berg).

Franklin, I.R. and R. Frankham (1998) 'How Large Must Populations Be to Retain Evolutionary Potential?' *Animal Conservation*, 1, 69–70.

Gautschi, B., et al. (2003) 'Effective Number of Breeders and Maintenance of Genetic Diversity in the Captive Bearded Vulture Population', *Heredity*, 91, 9–16.

Goudie, A. (2013) *The Human Impact on the Natural Environment* (Oxford: Wiley-Blackwell).

Haraway, D. (2003) *Companion Species Manifesto* (Chicago: Prickly Paradigm Press).

Hayles, K. (1999) *How We Became Posthuman: Virtual Bodies in Cybernetics, Literature, and Informatics* (Chicago: University of Chicago Press).

Kaiser, M. (2005) 'Assessing Ethics and Animal Welfare in Animal Biotechnology for Farm Production', *Revue Scientifique et Technique Office International des Epizooties*, 24(1).

Marker, L. and S. O'Brian (1989) 'Captive Breeding of the Cheetah (*Acinonyx Jubatus*) in North American Zoos (1871–1986)', *Zoo Biology*, 8(1): 3–16.

McPhee, E. (2004) 'Generations in Captivity Increases Behavioral Variance: Considerations for Captive Breeding and Reintroduction Programs', *Biological Conservation*, 115(1): 71–77.

Mullan, B. and G. Marvin (1999) *Zoo Culture* (Chicago: University of Illinois Press).

Nussbaum, M. (2004) 'Beyond "Compassion and Humanity": Justice for Nonhuman Animals', in C. Sunstein and M. Nussbaum (eds) *Animal Rights: Current Debates and New Directions* (Oxford: Oxford University Press).
Reiss, M.J. and R. Straughnan (1996) *Improving Nature: The Science and Ethics of Genetic Engineering* (Cambridge: Cambridge University Press).
Rollin, B. (1995) *The Frankenstein Syndrome: Ethical and Social Issues in the Genetic Engineering of Animals* (Cambridge: Cambridge University Press).
Rollin, B. (2003) 'Ethics and Species Integrity', *American Journal of Bioethics*, 3(3).
Rollin, B. (2006) *Science and Ethics* (Cambridge: Cambridge University Press).
Rolston, H. (1993) *Environmental Ethics* (Philadelphia: Temple University Press).
Rudy, K. (2011) *Loving Animals* (Minneapolis: University of Minnesota Press).
Ryder, O. and K. Benirschke (1997) 'The Potential Use of Cloning in the Conservation Effort', *Zoo Biology*, 16: 295–300.
Sagoff, M. (2001) 'Biotechnology and the Natural', *Philosophy and Public Policy Quarterly*, 21.
Siipi, H. (2008) 'Dimensions of Naturalness', *Ethics and the Environment*, 13(1): 71–103.
Smutts, B. (1999) 'Reflections' in J.M. Coetzee (ed.) *The Lives of Animals* (Princeton, NJ: Princeton University Press).
Snyder, N., S. Derrickson, S. Beissinger, J. Wiley, T. Smith, W. Toone and B. Miller (2002) 'Limitations of Captive Breeding in Endangered Species Recovery', *Conservation Biology*, 10(2): 338–48.
Spotte, S. (2006) *Zoos in Postmodernism: Signs and Simulation* (Madison, NJ: Fairleigh Dickinson University Press).
Thompson, P.B. (2007) *Food Biotechnology in Ethical Perspective*, 2nd edn (Dordrecht: Springer).
Turner, S. (2002) 'Jurassic Park Technology in the Bioinformatics Economy: How Cloning Narratives Negotiate the Telos of DNA', *American Literature*, 74(4): 887–909.
Verhoog, H. (2003) 'Naturalness and the Genetic Modification of Animals', *Trends in Biotechnology*, 21(7): 294–7.
Warkentin, T. (2009) 'Dis/Integrating Animals: Ethical Dimensions of the Genetic Engineering of Animals for Human Consumption' in C. Gigliotti (ed.) Leonardo's Choice: Genetic Technologies and Animals (Dordrecht: Springer).
Warren, M.A. (2002) 'The Moral Status of the Gene' in Justine Burley and John Harris (eds) *A Companion to Genethics* (Oxford: Blackwell Publishing).

6
From Protection to Restoration: A Matter of Responsible Precaution

Anne I. Myhr and Bjørn K. Myskja

Introduction

Climate change and human activities are the main reasons for the increasing degradation of ecological systems. Advances within modern biotechnologies, as genetic engineering and synthetic biology, hold promises for saving endangered species and revival of extinct species. The same biotechnologies are followed with concerns for unforeseen environmental harm. This has resulted in a call for regulation, including the application of the precautionary principle. There are, however, significant disagreements on how to apply the precautionary principle and how to handle risk. A key question concerns the meaning of the precautionary demand to avoid 'morally unacceptable harm' to the environment (UNESCO COMEST 2005) when applying new technologies. This needs further elaboration and we will draw some lessons from the controversies surrounding the introduction of genetically modified organisms (GMOs). Some similarities are for instance related to how to acknowledge risk and uncertainties, the existence of systemic properties of ecosystems and how to study the effect of changes induced in ecosystems. It has been argued that risk claims and awareness to uncertainties depend on assumptions about which facts and values are to be considered and which are to be left out. This may have implications for what we consider to be environmental harm as well as how we regard the significance of environmental harm that is identified.

Emerging biotechnologies may change the relationship between humans and nature (Kaebnick 2009; ECHN 2010; Gutmann 2011). In general the relationship between humans and nature is a challenging issue and there are different values and ideologies behind the various conceptual approaches. Within approaches used to ensure environmental and biodiversity protection, broadly defined economic approaches and terminology are increasingly employed. This introduction of economics into environmental and biodiversity issues, which has been adopted in concepts such as ecosystem services, has been challenged. The language of economics is also used in the assessment of new technologies, for example in cost–benefit analyses. We hold that there is a significant difference between non-monetary and monetary values approaches, with implications for the restoration of ecosystems and for the re-creation of species. Furthermore, an ecosystem management approach that is based on a more fundamental understanding of ecosystem functioning and interactions will be useful both for approaches that aim to protect and for those that are initiated for restoring ecosystems. We also question whether a precautionary approach is adequate when we seek to take into account crucial aspects of the ecological complexity that will be intervened with. Rather than a narrow interpretation of the precautionary principle with emphasis on 'avoidance of harm', we will argue that the focus should be on an ethics of responsibility approach.

This normative approach is influenced and derived from Hans Jonas's (1979) claim that the development of technology has given humankind the power to destroy the possibility of any future human life on earth, which leads to an imperative of responsibility: 'act in a way that makes the consequences of your acts conform with the permanent real human life on earth' (ibid., p. 36). Here Jonas points out that this responsibility is shared between humans since choices on issues of the development and the use of technologies are collective acts. An implicit part of this imperative of responsibility is that there is a duty to acquire adequate knowledge in order to contribute to the good of future generations, which implies the acceptance of uncertainty (ibid., p. 28), while at the same time encouraging innovation.

We will therefore argue that there is a need for a combined scientific and ethical analysis that involves the concept of responsibility to provide a sufficient basis for decision-making.

Use of modern biotechnologies for restoration purposes

Advances in molecular biology as genetic engineering techniques and synthetic biology applications hold the promise of their being used both to save endangered species and to revive extinct species. To save endangered species includes for example cloning, interspecies nuclear transfer and the production of stem cells. In California stem cells have been produced that are similar to those found in early embryos using cryopreserved cells collected from critically endangered animals such as rhinoceroses and monkeys (Callaway 2011), while in Australia the same approach has been used on the snow leopard. The scientists behind these projects aim to convert these stem cells into germ cells that could diversify the gene pools of threatened species (Kumar 2012). There are, according to Kumar (2012, p. 9), initiatives to 'collect as many DNA samples from endangered, threatened and extinct species as we can, so that if the human population ever reduces its footprint on Earth, these species can be reintroduced'. One example is the 'frozen zoo' at San Diego Zoo in California that has been maintained since 1976 and holds approximately 8,400 samples from more than 800 species and subspecies, including DNA, sperm, eggs and embryos, stored in liquid nitrogen. At present there are also other facilities around the world that have started collecting samples of frozen cells containing DNA from endangered animals before they go extinct and which is seen as an insurance (Kumar 2012). Most probably, this is for the purposes of a later use of modern biotechnology for reviving these species in case they are directly threatened by extinction. However, there is no reason to wait until this point. If the technology is successful, it would be better to make the species more robust now so as to keep it off the red list.[1] This is valuable for maintaining the ecological niche of the species and to minimize human intervention.

Another approach concerns the revival of extinct species. There is a huge interest in the revival of extinct species for several reasons. From a scientific viewpoint revival is especially relevant for species that have unusual traits, where it would be of high interest to study behavioural and physiological characteristics. Some would also argue it is valuable for restoration purposes and tourism (Kumar 2012). The revival of extinct species could also be pursued for acquiring information about species' evolutionary patterns, which also opens up the possibility of bioprospecting for interesting genes that could be identified and used for the expression of bioactive compounds of relevance in human and veterinary medicine (Kumar 2012). Collaboration between different scientific disciplines, from synthetic biology and biological conservation, could facilitate both the revival of extinct species and the protection of endangered species (Redford et al. 2013). In addition, many hold that the non-human world has intrinsic value (Rolston 1982) and that we should not only minimize human intervention, but also even repair the damage of previous interventions.

Biocontrol for ecosystem management

Biological control (biocontrol) aims to control pests (including insects, mites, weeds and plant diseases) using other living organisms. Biocontrol is used in pest management strategies and also to combat insect-borne diseases (such as malaria). Genetic engineering techniques provide new possibilities, for instance releasing GM viruses that through infection distribute infertility into pest organisms. Another approach is genetically modifying existing organisms in order to reduce their environmental harm. One example is the daughterless carps (Gilna et al. 2013) enabling control over an invasive species. One other potential way to use this technology includes the protection of specific species, as, for example, the wild salmon stock from the threat from escaped farmed salmon. Although wild salmon as such is not a threatened species, their genetic diversity arguably is under pressure from several sources, including escaped farm salmon that through mating with wild salmon can change their genetics or by competition reduce the number of wild salmon.

Synthetic biology and the creation of species

Synthetic biology has been defined as 'a scientific discipline that relies on chemically synthesized DNA, along with standardized and automatable processes, to address human needs by the creation of organisms with novel or enhanced characteristics or traits' (Presidential Commission for the Study of Bioethical Issues 2010).

There are different approaches that geneticists can use to associate various genes and gene variants with a range of traits in organisms (Collins 2012). In the case of simple metabolic or synthetases, it is even possible to combine individual gene sequences into genomic DNA, almost like genetic 'lego' blocks. Genetic engineers have developed different methods to alter the genome of living organisms. Synthetic biology differs from genetic modification in its scope, capacity for genetic changes and source material. Recent increases in the capacity of DNA synthesis technologies has made it possible to generate whole genes as well as genetic fragments that can be spliced together with methods already established within genetic engineering. This capacity makes it possible to build genes along with other functional elements. Rather than modifying an already existent DNA strand, synthetic biology builds the genetic fragments or genes together from a bottom-up approach into a synthesized genetic material that can be placed into a living cell which will have gained a number of specific novel functions. Currently, synthetic biology has mainly been done in microbes, because of their small genomes and ease of transformation. In the future, synthetic biology may be used both to re-create extinct species and to create new species.

Environmental risks by application of modern biotechnologies

New possibilities with modern biotechnologies where the products are intended for environmental release may represent new risks.

Analogy with GMOs

The environmental uncertainty introduced by GMOs is related to the intended benefits as well as the potential harms to ecosystems and social systems. Ecosystems are complex and it is difficult or, in some cases, impossible to interpolate from the laboratory and controlled field studies what will actually happen after introduction and release into the environment. Environmental effects of GMO use and release may arise due to interactions between the introduced transgene(s) and the recipient genome, or unanticipated interactions between the GMO and the ecological system. From the start, many ecologists were questioning the assumption that the large-scale behaviour of GMOs can be extrapolated from effects studied in small-scale models (Wolfenbarger and Phifer 2000). Later reports have shown that there is a spread of genes from, for example, herbicide tolerant GM crops, causing problems as contamination of neighbouring crops and wild relatives developing into weeds. The issue of non-target effects is still a challenging issue; many laboratory studies have been performed to measure the effects on target and non-target insects by for example insect resistant (Bt) crops. While most of these studies have reported no indications for non-target effects of Bt crops, there are some that do show that there may be adverse effects. This has caused debates about the quality of the different models and with regard to the statistics used. With GM crops there is also an issue that growth conditions are geographically and climatically different, making it difficult to identify the cause–effect relationships between a GMO crop and its environment.

The types of uncertainties surrounding GMOs can be divided into three broad classes (Nielsen and Myhr 2007):

1. Reducible uncertainty, due to lack of knowledge and the novelty of the activity that can be addressed with more research and focused collection of empirical data.

2. Irreducible uncertainty, due to inherent randomness, variability and complexity in the biological system under consideration.
3. Uncertainty arising from ignorance given that the prevailing operating paradigms and models do not adequately represent the biological system evaluated.

It has also been argued that uncertainty about environmental impacts should not be understood as lack of scientific knowledge, but also as lack of coherence between competing scientific disciplines in the choice of models and basic assumptions (Sarewitz 2004). This has been an issue with GMOs.

Important lessons from GMOs include the necessity to analyse the organisms in question within the specific ecological, biological or socioeconomic context that it is supposed to fill and the need for different scientific disciplines and various approaches for investigating risk and uncertainties, both for management and monitoring. This is an important lesson that must be taken into account with future GMOs, and especially those that are to be released into the environment for biocontrol purposes. Furthermore, these organisms are designed for spread in the environment, which compared with GM crops adds to the uncertainty with regard to unforeseen environmental harm.

Environmental risks with synthetic organisms

As with GMOs, it must be expected that unintended effects may evolve with synthetic organisms. There are some similarities in the questions raised, for example with regard to how the synthetic organisms can persist and spread in the environment, and if inserted genes and altered DNA can be transferred to naturally occurring organisms. For example Dana et al. (2012) question whether synthetic microbes can disrupt the normal functions of ecosystems by transferring their altered DNA to other microbes. Such synthetic microbes can also be altered in more sophisticated and fundamental ways than GMOs, which may make it even more difficult to perform adequate risk-associated research in the laboratory; they may therefore also be more difficult to monitor in the environment.

When it comes to protection of endangered species, they have their niches in the environment, while extinct species have lost theirs in the ecosystems. Hence with the reintroduction of extinct species, their revival, in addition to raising concerns about unexpected effects due to the use of modern biotechnology, raises the question as to whether these organisms may increase competition for resources, or disrupt crucial ecological functions (ibid.). Here it is relevant to draw an analogy with the impacts made by invasive species. Most invasive species have not caused any problems and have been considered as an acceptable broadening of biodiversity. However, some species cause problems, such as the king crab along the Norwegian coast and rodents and rabbits in Australia, with the consequence that they have wiped out other species. It has therefore been argued that conservation and synthetic biologists need to start to discuss both the possibilities and concerns so as to be able to acknowledge better the impacts made by synthetic organisms on ecosystems, including protected areas (Redford et al. 2013).

The concerns for environmental risk were the reason for an assembly in July 2011 by the Synthetic Biology Project at the Woodrow Wilson International Center for Scholars in Washington DC. A group of synthetic biologists and ecologists gathered for discussions on how to assess risk and identify what the possible risks of introducing synthetic organisms into the environment could be.[2] One of the outcomes of this meeting was an agreement for an innovative and collective approach for the development of an eco-risk research agenda.

Environmental protection: do only economics matter?

Within environmental and biodiversity protection, broadly defined economic approaches and terminology are increasingly employed. The implicit values of such economics-derived methods, where all values arguably can be made commensurable by being given a specific price, may significantly influence policy and management approaches. For example, one can argue according to the logic of economics that ecosystem services may not necessarily require that the native species is introduced in order to restore an environment since GM and synthetic organisms may provide the required services. This can be illustrated using the potential loss of wild salmon in Norwegian rivers as an example (Myhr and Myskja 2011). Assuming improved fitness of farmed salmon (including, in the future, GM salmon), some would say that the escape of salmon with subsequent depletion of the wild stock due to interbreeding and competition represents no harm. We still have wild salmon, although somewhat different from the original variety. Others will, presuming that values are non-commensurable, argue that there are several reasons to reject the assumption that farmed salmon or hybrids can replace wild salmon without significant loss. They consider this to represent serious harm since:

1. The genetic interaction will result in reduced genetic diversity (although this may be contested).

2. The escaped farm salmon will replace the wild relative, which is problematic for our distinctions between nature and culture, expressed in the value placed by most people on the idea of naturalness. In this particular sense, the escapee is less natural than the native salmon.

Although the use of modern biotechnologies has achieved little attention among those who are concerned about loss of biodiversity, the example of wild salmon in Norwegian rivers illustrates an important question: how will organisms revived or saved by the use of modern biotechnologies be met in the ecosystems, and will they be accepted by game fishermen and society? Many will probably consider their replacement or reintroduction to be a serious harm.

Clearly we cannot determine what is a beneficial or what is a serious interference with nature, and certainly not what is to count as acceptable or unacceptable environmental harm, unless we determine which aspects of nature and its functioning is to be regarded as worth preserving and restored as it is today (Arntzen 2001). This is a matter of deciding which values should be taken into account and according to what kind of approach. Thus the question of environmental harm is correlated to conceptions of how to maintain and preserve biodiversity and how to restore ecosystems.

Cultural significance

We should not base the assessment of the value of wild nature solely on economics, although the economic aspects are essential as long as we accept an anthropocentric approach to value. If humankind is to survive, future people need sufficient resources to sustain life. This is the basis for the principle of sustainability. However, future humans will probably share some of our other value judgements concerning nature. These are not shared values in the sense that everybody accepts them or believes they should carry any weight in practical deliberation. They are shared in the sense that they serve as important premises in public debates concerning human interaction with nature, and that there are good arguments supporting them. We can distinguish at least three interconnected non-economic values of nature.

First, we may appreciate nature for its own sake, as something other than the social world of human beings. The mere fact that there is something we do not understand completely, something that we do not control, and something that is and will remain unexplored, makes it valuable. This idea connects to the concept of 'untouched nature' which usually has positive connotations, indicating that people generally understand such nature as valuable.

A second related cultural value is based on aesthetic appreciation of nature. Experiencing nature gives a pleasure that is not connected to any desire or cognitive interest-based evaluation as Kant (1987 [1790]) famously wrote. This disinterested pleasure is aesthetic pleasure connected to beauty or sublimity. In Kant's thinking, this experience is the link between theoretical knowledge and morality, and as such is an essential part of human life. It is important to note that the aesthetic judgement serves this key role just because it is disinterested. Thus, we have an interest in nature because it is valuable for us, even when we disregard any interest we may have in it.

Third, nature has recreational value for many. People appreciate nature because it enables them to disengage from their usual activities and experience something different. This recreational value may take many forms, from a merely aesthetic engagement, via nature as a place for physical exercise and challenges, to a direct involvement in live nature through fishing, hunting and gathering. The latter activities have an economic side for some, but for modern people this is not essential, as is demonstrated in catch and release fishing.

We may translate these values into economics as important elements in the tourist industry, but they cannot be reduced to this economic value. Even if there were no tourism or anybody making money from people's non-economic evaluation of nature, people would still appreciate nature for these reasons. There are no grounds for assuming this would be different in the future, so we must assume that people will appreciate nature for its own sake, as a source of aesthetic pleasure, and for recreation. Considering the increasing population and the ongoing decimation of nature, it is likely that these values will be even more important in the future. Protecting nature from human intervention will be important beyond the demand of ensuring a sound resource base for future generations.

In this perspective, protecting species and reintroducing extinct ones may make sense, but we cannot protect or reintroduce every species for capacity reasons. Choosing which ones we should allow can be decided on a number of arguments, including those referring to cultural values. Some species are more significant than others because they connect more directly to these ways of valuing nature. There are certainly strong cultural elements here, and no absolute, context-independent basis for appreciating one more than the other, though that is irrelevant in this context. It is a fact that we find some species more valuable than others because they play a more important role in our appreciation of nature, and that gives us grounds for special protection of these species. In some cases, the reason for this judgement is the key role played by a species in some particular ecosystem; at other times it is a key role played in a cultural community.

We have reasons to do something special to protect or even revive such iconic species, but not all that is desirable is morally good. We must find a way to decide whether using biotechnology for such purposes is acceptable, and whether the environmental harm is essential.

Social and ethical issues

Closely connected to cultural values we find ethical issues, although they tend to play a stronger normative role. We can draw lessons from the GMO debate, a controversy that arguably was more a matter of ethics and social issues than of scientific questions (de Melo-Martin and Meghani 2008). Several interpretations exist concerning what went wrong in the failed attempt to place GM crops on the European market (Craye n.d.). This has been given special attention with regard to the introduction of nanotechnology, though some central issues are equally important when considering the use of modern biotechnologies for preserving and protecting species and ecosystems. The troubling issue is the unexpected public scepticism to GMOs in Europe, which may signify that the scientific and political communities have failed to understand what is at stake in issues that may have significant environmental impacts.

Different interpretations emphasize various reasons for the resulting public rejection of GMOs in Europe, which has implications as to how to avoid related problems. According to a first interpretation, to avoid public scepticism the emphasis should be on anticipating possible negative technological risk impacts. According to a second interpretation, emphasis should be on a wider set of issues, such as the social, distributional, desirability of innovations, and cultural and ethical challenges, including debate about appropriate procedures for governance of emerging technologies (Craye n.d.). Whereas the first interpretation largely preserves the primacy of innovation and the basics of risk-focused decision-making (perhaps with an added 'precautionary element'), the second is based on a need for fundamentally transforming the 'modern' model of science-policy relations, following the insight that emerging technologies and social orders are co-produced. This means that the knowledge gained through science and technology are embedded in social norms, institutions and practices, while this techno-scientific knowledge at the same time contributes in shaping these practices. Scientific and societal developments are interwoven to a degree that there are reasons to regard the second interpretation to be more in keeping with the empirical realities of modern technological democracies (Nydal et al. 2012).

Thus, the ethical issues at stake in biotechnological protection or restoration of species are not merely a matter of calculating the risk, but of asking what kind of technological development will contribute to a better society. This is a matter of taking responsibility for the wider implications of the technological development. Under this model, one cannot assume that it is possible to outsource the ethical questions to ethical experts or to a political process after it is developed and ready to use. Without becoming technologically deterministic, it is plausible that a developed technological solution is difficult to reject – even if it is potentially harmful – when compared to adjusting plans at early stages of development. It is also true that the scientists doing the research are crucial for predicting the potential benefits and harms of a proposed technological solution.

In short, ethical and social issues must be part of the technology development from the beginning. But how should this ethical analysis take place, and who should be involved in it? One solution could be to integrate a precautionary approach in the development of the technology, as has been widely suggested within the GM debate.

Precautionary approach

The complexity of the natural and social systems involved implies that the information acquired in quantified risk assessments may be inadequate for evidence-based decisions. One controversial strategy for dealing with this kind of uncertainty is the precautionary principle, which has been a central element in the bio- and nanotechnology debates. Implementation of the principle entails two interrelated proposals: the first one concerns caution in the face of the application of new technology; the other emphasizes the importance of conducting risk-associated research. There is still a wide discussion of how to operationalize this principle or approach. This debate became especially significant in Europe where there was strong public resistance to the use of GMOs. One recent version of the principle highlights several crucial elements involved in it:

> When human activities may lead to morally unacceptable harm that is scientifically plausible but uncertain, actions shall be taken to avoid

> or diminish that harm. Morally unacceptable harm refers to harm to humans or the environment that is
>
> - threatening to human life or health, or
> - serious and effectively irreversible, or
> - inequitable to present or future generations, or
> - imposed without adequate consideration of the human rights of those affected.
>
> The judgment of plausibility should be grounded in scientific analysis. Analysis should be ongoing so that chosen actions are subject to review. Uncertainty may apply to, but need not be limited to, causality or the bounds of the possible harm.
>
> Actions are interventions that are undertaken before harm occurs that seek to avoid or diminish the harm. Actions should be chosen that are proportional to the seriousness of the potential harm, with consideration of their positive and negative consequences, and with an assessment of the moral implications of both action and inaction. The choice of action should be the result of a participatory process. (UNESCO COMEST 2005)

As we can see, there must be some plausible harm scenario and lack of sufficient information for generally acceptable risk assessments. However, even these rather precise criteria require interpretation and are subject to controversy. Risk assessments usually contain some subjective elements in addition to what are generally accepted facts. In new knowledge areas, it is reasonable that there is disagreement as to what can be counted as facts or as subjective elements. This became very clear in the GM debate, where almost every claim became subject to dispute. The fundamental disagreement concerned whether there was a need for precaution, as some thought this was a well-established technology but others connected it to a wide number of plausible harmful scenarios ranging from serious health problems to devastating consequences for biodiversity and food safety. Even those who agreed that precautionary measures were required failed to reach agreement on how to understand 'precaution' in this case. However, we can find one consistent interpretation of the principle in the Norwegian regulatory regime, where every application for approval of a specific GMO was sent back to the producer with demands for an analysis of the extent to which that variety was sustainable and beneficial for society. Some regard this practice as being unreasonably averse to technology, and it would be impossible to implement in any country with a strong biotechnology industry.

The continued controversy concerning the precautionary principle demonstrates that it is problematic (Myhr and Myskja 2011). Even if it is a good tool for shifting the burden of proof from the critics to the technology proponents, it is not clear that such shifting of the burden always is desirable or justified. Furthermore, it is not clear what 'precaution' means as long as it is not equivalent to a moratorium on the proposed activity. In addition, there are moral problems connected to the demand of participatory processes guiding action. The UNESCO document specifies this as public participation. We agree that there is a need to open up, and that social groups need to be included for discussions of the uncertainties involved. Marris and Rose (2012, p. 29) do for example stress that this is especially important with synthetic biology to ensure that 'different visions of what is desirable are debated before particular ones become entrenched and hard to modify'. However, we are concerned as to whether it is possible to recruit lay people to such time-consuming and demanding tasks, assuming that they have to go through a challenging learning process. Even if one manages to recruit them, their capacity for contributing may be limited. One would assume that regulating the use of a complex and advanced technology in a complex environment is a matter for experts. We are not saying that lay people lack the capacity for contributing to this debate, but accepting the task means taking responsibility for acquiring sufficient knowledge. How would they know when they were sufficiently informed, so that, when the decision was made, it became their responsibility?

There are studies showing that lay people are willing to spend time on participating in engagement studies, but the role assigned them in precautionary deliberation seems more demanding than many regular exercises where people are asked about matters that directly affect their everyday lives. In these cases, their particular experiences and work lives provide unique perspectives to the debate. Recruitment to discussions of issues that are more general and require specialist knowledge is difficult, and there are indications that many would prefer these complex issues to be handled by experts (Skolbekken et al. 2005). Even if we disregard these problems and assume that lay people willingly participate and do the job of gaining the knowledge required for well-founded participation, this implies a shift of responsibility from the experts and the political class – which is morally problematic. If using biotechnology in order to protect endangered species or re-create extinct ones may lead to scientific uncertainty concerning environmental harm, it is not clear that the precautionary principle is the answer.

Protection and restoration: a matter of responsibility

Hans Jonas (1979, p. 36) suggests a new categorical imperative to replace the Kantian one: 'act in a way that makes the consequences of your acts conform with the permanent real human life on earth'. The context giving rise to this imperative is the power humankind has acquired through technology, that in the worst case may lead to the destruction of the possibility of any future human life on earth (ibid., pp. 26f.).

In addition, Jonas says that this new[3] context also alters who is subject to the imperative. We all share responsibility, since our acts are collective acts. This does not necessarily imply that everyone is equally responsible, but no one is responsible *alone* for deciding to develop and use different technologies. The moral imperative becomes political, and being citizens of a technology-driven society, who have knowledge and possibilities of influence, we share responsibility for these decisions. Our responsibility for future generations must be at the core of any precautionary thinking: we should leave them an Earth that gives them the opportunity to decide what kind of life they want to lead. Precautionary thinking and the imperative of responsibility mean that the Earth should not be left in a worse state than that which we received. This implies a special responsibility to combat and reduce biodiversity loss.

Jonas's principle of responsibility is compatible with the precautionary approach but has wider scope. He explicates our duty to gain knowledge in order to contribute to the good of future generations, which implies the acceptance of uncertainty (ibid. 28), while at the same time encouraging innovation. Many have argued that a restrictive precautionary policy may prevent rather than promote optimal conditions for future people (Sunstein 2002) and we have argued that it may redistribute responsibility in an unacceptable way. We think that an ethics of responsibility approach that is directed to how we should promote a good life for future generations is a better way to ensure a sustainable development of technology.

It is challenging to operationalize the idea of responsibility in protection of threatened species or for the restoration of extinct ones. Despite the collective nature of responsibility for technology development and use, the imperative of responsibility is subjective in the sense of the Kantian categorical imperative. This means that each one must interpret how to apply the principle in one's own life. The interpretation will be dependent upon education, work, social status and abilities. This means that those with scientific knowledge or power to influence technology development carry more responsibility for the decisions on how to apply technologies than representatives of the public. However, lay people are also obliged to get involved in gaining knowledge in order to take part in public discussions on how the future of our techno-society should be. Those who are involved in technology development or have knowledge about the context for its use still have a special role to play in these discussions to ensure that they are based on the best available knowledge.

There is a significant moral difference between protection and restoration from a responsibility perspective. We are primarily responsible for our own actions, including our omissions. When we consider our collective responsibility for ensuring good living conditions on a future earth, leaving as much of nature as free from human intervention as possible appears to be a good rule of thumb. This ensures the largest possible resource base for future humans, as well as their possibility to appreciate the cultural values of nature in the same way as we can. The problem is that the sheer number of humans and the powerful technologies we have developed means that our activities always affect nature and no part of it can remain untouched. We need to take counteraction against these interventions in order to fulfil the imperative. In this context, there is need for some degree of precaution.

When we intervene in nature in order to protect species threatened by human activity, our action may be either a way to take responsibility or be expressive of technological quick-fix thinking. This depends on the number of alternatives we have. Using biotechnology to protect an endangered species is similar to saving it from extinction by taking the remaining individuals into human custody. If there are no alternatives that are more in keeping with leaving nature to itself, it is better than doing nothing. Even if there is unforeseen harm, or the attempt fails, we may have no better options. As long as the species has a niche in the ecosystem, it may be morally right to intervene to protect it. However, the condition is that we have reasons to consider this species sufficiently valuable to accept the risks and uncertainties concerning the intervention.

Re-creating extinct species for restoration purposes is a different matter, morally speaking. Here the ecosystem niche has disappeared with the animal, and it is in principle impossible to reintroduce it to the same ecosystem. This means that the restoration, ecologically speaking, is similar to the introduction of a foreign species, with all the uncertainties that follow from such interventions that draw on experiences that conservation biologists have had with invasive species. We know from previous cases that the consequences of such introductions may range from beneficial to devastating. In such cases, there must exist good reasons for us to say that the intervention is acceptable. The interests of technology development or scientific curiosity are not sufficient grounds for accepting this degree of scientific uncertainty.

Notes

1. See www.iucnredlist.org.
2. See http://www.synbioproject.org/events/archive/cea/.
3. The novelty of the situation is relative. The new and powerful technologies have been with us for almost a century, but our moral and political response is still based on intuitions belonging to earlier times.

References

Arntzen, S. (2001) 'Integrity and Uses of Nature', *Global Bioethics: Problemi di bioetica*, 14, 67–75.

Callaway, E. (2011) 'Could Stem Cells Rescue an Endangered Species?', *Nature News*, 4 September.

Collins, J. (2012) 'Synthetic Biology: Bits and Pieces Come to Life', *Nature*, 483, S8-S10.

Craye, M. (n.d.) 'Governance of Nanotechnologies: Learning from Past Experiences with Risk and Innovative Technologies', unpublished manuscript.

Dana, G.V. et al. (2012) 'Synthetic Biology: Four Steps to Avoid a Synthetic-Biology Disaster', *Nature*, 483(29).

ECNH (Federal Ethics Committee on Non-Human Biotechnology) (2010) *Synthetic Biology: Ethical Considerations*, www.ekah.admin.ch/fileadmin/ekah-dateien/dokumentation/publikationen/e-Sy nthetische_Bio_Broschuere.pdf (accessed 14 December 2012).

Gilna, B. et al. (2013) 'Governance of Genetic Biocontrol Technologies for Invasive Fish', *Biological Invasions*,

Gutmann, A. (2011) 'The Ethics of Synthetic Biology: Guiding Principles for Emerging Technologies', *Hastings Center Report*, 41(4), 17–22.

Jonas, H. (1979) *Das Prinzip Verantwortung* (Frankfurt am Main: Suhrkamp Verlag).

Kaebnick, G. (2009) 'Should Moral Objections to Synthetic Biology Affect Public Policy?', *Nature Biotechnology*, 27(12).

Kant, I. (1987 [1790]) *Critique of Judgment* (Indianapolis: Hackett).

Kumar, S. (2012) 'Extinction Need Not Be Forever', *Nature*, 492(9).

Marris, C. and N. Rose (2012) 'Let's Get Real on Synthetic Biology', *New Scientist*, 11 June.

Melo-Martin, I. de and Z. Meghani (2008) 'Beyond Risk: A More Realistic Risk-Benefit Analysis of Agricultural Biotechnologies', *EMBO Reports*, 9, 302–6.

Myhr, A.I. and B.K. Myskja (2011) 'Precaution or Integrated Responsibility Approach to Nanovaccines in Fish Farming? A Critical Appraisal of the UNESCO Precautionary Principle', *Nanoethics*, 5, 73–86.

Nielsen, K.M. and A. Myhr (2007) 'Understanding the Uncertainties Arising from Technological Interventions in Complex Biological Systems: The Case of GMOs', in T. Traavik and L. Lin (eds) *Biosafety First: Holistic Approaches to Risk and Uncertainty in Genetic Engineering and Genetically Modified Organisms* (Trondheim, Norway: Tapir Academic Press), 108–22.

Nydal, T et al. (2012) *Nanoethos*. Report to the ELSA programme in the Research Council of Norway, Oslo.

Presidential Commission for the Study of Bioethical Issues (2010) *New Directions: The Ethics of Synthetic Biology and Emerging Technologies*, http://bioethics.gov/cms/sites/default/f iles/PCSBI-Synthetic-Biology-Report-12.1 6.10.pdf (accessed 17 December 2012).

Redford, K.H et al. (2013) 'Synthetic Biology and Conservation of Nature: Wicked Problems and Wicked Solutions', *PLoS Biology*, 11(4).

Rolston, H. III (1982) 'Are Values in Nature Subjective or Objective?', *Environmental Ethics*, 4(2), 125–51.

Sarewitz, D. (2004) 'How Science Makes Environmental Controversies Worse', *Environmental Science and Policy*, 7, 385–403.

Skolbekken, J.-A et al. (2005) 'Not Worth the Paper it's Written on? Informed Consent and Biobank Research in a Norwegian Context', *Critical Public Health*, 15(4), 335–47.

Sunstein, C.R. (2002) 'The Paralyzing Principle', *Regulation*, 25(4): 32–7.

UNESCO COMEST (2005) *Report of the Expert Group on the Precautionary Principle of the World Commission on the Ethics of Scientific Knowledge and Technology (COMEST)*, http://unesdoc.unesco.org/images/0013/001395/139578e.pdf (accessed 22 August 2013).

Wolfenbarger, L.L. and P.R. Phifer (2000) 'The Ecological Risks and Benefits of Genetically Engineered Plants', *Science*, 290, 2088–93.

7
Just Fake It! Public Understanding of Ecological Restoration

Bart Gremmen

Introduction

When you travel by train from Amsterdam to Lelystad, you do not expect to see many herds of wild animals in an open landscape. What you do see, however, are herds of Heck cattle and Konik horses, and huge herds of red deer! This is possible in the nature reserve of the Oostvaardersplassen. Some people even compare it to areas in Africa like the Serengeti (Vera 2009). The development of this area marked the beginning of an interesting switch in Dutch nature conservation policy from a defensive approach to the development of a new nature, more proactive approach. For more than one hundred years, people had tried to preserve the traditional agrarian cultural landscape, which had to be protected against industrialization, recreation and the expansion of cities. This resulted in the development of a preservationist movement, which tried to save nature by buying land. In the 1960s, in line with scientific progress in ecology, ecological engineers started to restore areas and 'bring nature back'. Two major Dutch players were responsible for this switch: the National Forest Service and the Society of Nature Conservancy. The Oostvaardersplassen lakes area and the river delta area of the Millingerwaard are the first Dutch areas where large herbivores have been introduced into nature reserves.

Around 1980, ecological engineers, notably in the USA, argued that the destruction of nature in a certain area, for example if the use of the area had been changed to industrial, could be repaired by the ecological restoration of nature. This resulted in a clash in environmental ethics between the protectionist and the ecological restoration perspectives. The discussion started when Robert Elliot, from the protectionists' side, claimed that ecological restoration could only result in 'fake nature' (Elliot 1982). Even today, this discussion is continuing (Gamborg et al. 2010). Thirty years ago in the Netherlands, many nature conservationists, some members of the media, a few farmers and some of the general public claimed that the development of so-called 'new nature' could only result in 'fake nature' (Koene and Gremmen 2002). They all opposed the development of new nature reserves by ecological engineers who claimed to be following the main principles of ecological restoration. They became suspicious because these developers were telling stories about how in ecologically restored areas in the Netherlands aurochs had returned to the wild; yet at the same time they were hearing reports from the media that these aurochs were in fact artificial reconstructions, that is look-alikes. The media were also reporting that local residents were complaining about open wounds and hunger in animals, as well as flooding. In winter, farmers were trying to feed these wild cattle and horses. In a country with only manmade landscapes, these new nature reserves could at best be considered as landscapes abandoned or neglected after their creation.

There have been many more ecological restoration projects in the Netherlands since the 1980s, almost all surrounded by their own scientific, political and public debates. These debates were often about the value of 'new nature'. I guess that Elliot's verdict would have been that the Dutch policy switch to ecological restoration has only resulted in fake nature. This is understandable from the perspective of the protection of existing (primeval) nature. However, as I will show later in two case studies, in the first examples of Dutch ecological restoration the local former nature was not restored. As a consequence, no fake nature has been developed, but a kind of new, and – apart from the animals introduced – spontaneous nature based on manmade conditions. In contrast to ecological restoration, I would like to call this the 'developing new nature' approach. Although the nature in this developing new nature approach is not fake, I want to demonstrate in this chapter that its developers have been *faking*. By linking the nature that has resulted from the developing new nature approach to the prehistoric period of aurochs and tarpans, they have presented this nature as 'the return of nature' by ecological restoration. This has not only severely damaged public trust in nature management but has also clouded the rise of the developing new nature approach as a third approach in addition to the protectionist and the ecological restoration approaches.

After a brief sketch of the debate on faking nature in environmental ethics, I will focus on the Oostvaardersplassen and on the Millingerwaard. I will compare and evaluate these areas on five principles of the ecological restoration approach: the naturalness of the original area; spontaneous nature; authentically wild animals; the shift from management to monitoring; and the shift from public debate to public communication.

The faking nature debate in environmental ethics

Human pressure on natural areas is still increasing. To ensure economic growth, we need ever more scarce land, gas, oil, and so on. On a global scale, government agencies and industries would like to start open-cast mining, the clear-felling of forests, and other technical interventions in nature reserves. As mentioned in the introduction, more than 30 years ago, from a protectionist perspective, Elliot started a debate that is still ongoing (Gamborg et al. 2010). He warned against technical interventions in nature reserves because, once we have destroyed nature, it is lost forever (Elliot 1982). We can replace it with something else, but we are unable to restore its original characteristics. In his view, the ecological restoration perspective could be used to justify nature destruction because it claims that we could restore nature, and even that we would not notice the difference from the original. Elliot claims that such restoration is just faking nature.

Elliot's target is the so-called 'restoration thesis' of ecological engineers. This is the idea that 'the destruction of something of value is compensated for by the later creation (recreation) of something of equal value' (ibid., p. 381). Elliot claims that this thesis is false. His 'anti-restoration thesis' involves highlighting and discussing analogies between faking art and faking nature. He argues that natural areas have values that artificial or restored ones lack and that our 'wilderness valuations depend in part on the presence of properties which cannot survive the disruption/restoration process' (ibid., p. 382). A restored area, then, since it cannot have such properties, cannot be as valuable as a natural area.

The main argument of Elliot's anti-restoration thesis is that genesis is a significant determinant of an area's value: it is a causal continuity with the past. The argument depends on the difference between artificial and natural genesis. Elliot's main concern is with ecological engineers who would use restoration to undermine preservationist arguments: if natural areas can be restored, why not use the resources on or under them, as and when it becomes necessary to do so, and restore them when we are finished with using them? His arguments, however, seem to apply equally to the restoration and reclamation of areas not used in this way. Why is the restoration thesis false?

Elliot offers three arguments. First, 'a [natural] area is valuable, partly, because it is a natural area; one that has not been modified by human hand, one that is undeveloped, unspoilt or unsullied' (ibid., p. 383). Now if you assume at the outset that the effects of modification by human hand are always bad (spoiling or sullying), then of course natural areas will be more valuable than ones modified by human hand. But why do we have to assume this? And why must we assume that not being modified by human hand is always a consideration in an area's favour?

Elliot's second argument is based on an analogy with the field of aesthetics: 'thus we might claim that what the ecological engineers are proposing is that we accept a fake or a forgery instead of the real thing. If the claim can be made good then perhaps an adequate response to restoration proposals is to point out that they merely fake nature: that they offer us something less than was taken away' (ibid.). Now, in most cases, a fake or forgery is less valuable than an original. But why think of the products of restoration as fakes or forgeries? A work of art often becomes more valuable after restoration. Also, forging and faking are species of deception; but restoring something is by definition not deceiving people, because there is no need to hide the restoration or lie about it.

Elliot's third argument is that natural areas are more valuable than restored areas because some people value natural areas – or areas with certain kinds of genesis or origin – more than restored ones. The basis for what Elliot says relies largely on his claim that 'what is significant about wilderness is its causal continuity with the past' (ibid., p. 385), and that through man's interference we cease this causal continuity and remove, or at least distance, nature from its valuable origins, However, just because some people value something, it does not follow that it is valuable or more valuable than something that some people do not value.

The most serious issue raised by Elliot's anti-restoration thesis is that 'natural' means something like 'unmodified by human activity'. The assumption here, however, is that human beings are not part of nature, and that human activity and its products are unnatural. This is a difficult assumption because, as humans, not only are we part of nature, but also we cannot help changing the world all the time. To do otherwise, we would have to stay outside nature, but a very long time ago our activities were already having global effects.

Elliot's claim is about existing, undisturbed nature. Katz expands Elliot's claim by stating that 'once we begin to create restored natural environments, we impose our anthropocentric purposes on areas that exist outside human society...they will be anthropocentrically designed human artifacts' (Katz, 1992, p. 98). Katz would like people not to 'misunderstand what we humans are doing when we attempt to restore or repair natural areas' (ibid., p.106); we can never re-create nature. We can only satisfy our anthropocentric desires by manipulating and dominating nature to create artefacts that merely appear natural. Maybe most people would prefer undisturbed nature over restored areas. However, restoration techniques represent a way of saving values that would otherwise be lost. In addition to undisturbed nature, restored nature could also be valuable. In my view, the development of new nature is a third alternative in which we let plants and animals co-evolve to a fit. By definition, this new nature has no history. I do not think that this is fake nature with no, or less, value. Is it rather just that it has a different value compared to undisturbed nature or restored nature?

Although new nature is not fake, I would like to raise the question of whether ecological engineers have deceived us by linking the new nature to the prehistoric period of aurochs and tarpans, and by presenting this new nature as 'the return of nature' by ecological restoration. I therefore make a conceptual switch from *fake* nature to the *faking* of nature, and, using the five principles of the ecological restoration approach, I focus on the normative dimension of faking in two cases of new-nature development.

Case I: the Oostvaardersplassen

The naturalness of the original area

The Oostvaardersplassen reserve is situated north-west of Amsterdam in the former Southern Sea. In 1918, after a flooding disaster in 1916, the Dutch government decided to turn the Southern Sea into a lake and to develop a series of polders. In 1968, the last polder, the South Flevoland polder, was reclaimed from Lake IJssel. The Oostvaardersplassen is that part of the 'new' land originally designated as an industrial area. Consequent to the worsening economic circumstances, the industrial area remained undeveloped, and the area was redesignated as a nature reserve. The Oostvaardersplassen consists of 6,000 ha.: about 3,600 ha. are open water and marshland, and about 2,000 ha. are wet and dry open grasslands.

How natural was the situation in 1968 at the beginning of the Oostvaardersplassen? In order to develop the area, the complete ecosystem at the bottom of that particular part of Lake IJssel was destroyed by human intervention. The resulting polder was never meant to become nature. Instead of protecting nature, human activities had prepared an industrial area, the opposite of a natural area. From a protectionist perspective, the Oostvaardersplassen is totally unnatural. From an ecological restoration perspective, only the original ecosystem of Lake IJssel could have been restored. The resulting situation, however, was new in every aspect. Because neither a protectionist nor an ecological restoration perspective fits the Oostvaardersplassen area, the developing new nature perspective is a more useful approach to adopt.

Spontaneous nature

As the bottom of the former Lake IJssel, the soil of the Oostvaardersplassen consists of fertile calcareous clay. Soon after the polder was ready, many water pools formed and the process of marsh development started. The National Forest Service website stresses the spontaneous way nature 'returned' to this area. After a while, a lot of different kinds of plants and animals took over the bare land. Bird species that had become very rare in the Netherlands, such as the marsh harrier, *Circus aeruginosus*, and the bearded tit, *Panuris biarmicus*, established themselves as breeding birds in numbers that were high in comparison to other nature reserves in north-western Europe (Vera 1988). Even animals that had disappeared as breeding species in the Netherlands returned, such as the white-tailed eagle, *Haliaeëtus albicilla*, a pair of which established a territory in the area and has bred since 2006. Also up to 30,000 (non-breeding) greylag geese, *Anser anser*, retreat to the marshes to moult their wing feathers (Van Eerden et al. 1997). In his reconstruction of the first phase of the Oostvaardersplassen, Vera emphasizes, in combination with fluctuations in the water level, the management role of these geese. They have created open marsh vegetation by grazing on the common reed, *Phragmites australis*, thus taking over the traditional human management of reed cutting and enabling many species to exist in the area (Vera 2009). The greylag geese were responsible for this change and 'have caused a shift in our thinking about the potential for creating conditions where ecosystems can function naturally' (Ibid., p. 32).

The standard story told is that the new nature of the Oostvaardersplassen returned in a spontaneous manner. That would have meant the adoption by the National Forest Service of a strict hands-off policy in their management of the area. However, this standard story is false. From the very start, many technological measures were taken. For example, in the wet areas, reed seedlings were dropped by small aeroplanes, and grass seed was thrown on the dry land. Ditches were closed and water holes were dug. In 1982, the railway from Almere to Lelystad was built around the area. The 'return' of the geese was just a short pause in their migration journeys.

Authentically wild animals

Without grassland, the geese in the Oostvaardersplassen would not be able to congregate before and after their moult; this in turn would end their role in the ecosystem. A possible solution was to incorporate farming into a dry area adjacent to the marsh and use domestic cattle to create and maintain open grasslands. Some scientists argued that, if this was so, their wild ancestors, the aurochs and the tarpans, must also have been able to do this (Vera 1988; Vulink and Van Eerden 2001). Because these wild ancestors had become extinct, in 1983 the National Forest Service chose 35 Heck cattle and 27 Konik horses as suitable replacements. The situation in the Oostvaardersplassen was interpreted as a prehistoric ecosystem. The situation of about 20,000 years ago was taken as the time of reference. Vera's main argument for using Heck cattle and Konik horses is 'because they have undergone very little selective breeding and may therefore have many of the characteristics of their wild ancestors' (Vera, 2009, p. 33). In 1992, the National Forest Service introduced 54 red deer to control the scrub.

The introduction of Heck cattle and Konik horses is not only part and parcel of ecological restoration, but is also a clear case of faking by the experts. Contrary to Vera's claim that Heck cattle have undergone very little selective breeding and may therefore have many of the characteristics of their wild ancestors, this breed of cattle was designed and developed by the brothers Heinz and Lutz Heck in prewar Germany. Both were directors of zoos, Heinz in Munich and Lutz in Berlin. They used the drawings in the French Altamira cave as their inspiration. They wanted to create an auroch look-alike. Lutz Heck combined Spanish and French fighting bulls in the 1940s. These animals all died in 1945. Heinz tried to create aurochs by combining landraces from Germany, Hungary, Scotland and Corsica. They weigh 600 kg and their height is 1.4 m. At first sight, with their big horns, they look like aurochs, but they are 400 kg and 30 cm too short. A few of these animals survived and were shipped to Poland where they were kept until Vera transported 35 Heck cattle in 1983 to the Oostvaardersplassen. Should we consider this as a back-breeding strategy in ecological restoration or are these animals just look-alikes?

The shift from management to monitoring

The process of biotic dynamics through releasing certain species of large herbivores plays a key role in the developing new nature approach. The conditions of this process are completely manmade and a good example of ecotechnology (Gremmen and Keulartz 1996). Because animals differ in what they eat and how they graze, they have a determinative influence on the development of particular areas. The introduction of cattle and horses made it possible to regain a relatively open area. In November 2012, the National Forest Service counted approximately 4,800 animals in the Oostvaardersplassen: 312 Heck cattle, 1,110 Konik horses and 3,350 red deer (Staatbosbeheer 2013). The ultimate goal of the National Forest Service is a self-sufficient population. Management has to develop into minimal monitoring, the key to their hands-off policy. Everyday, the rangers keep an eye on the health and welfare of the large herbivores. This is comparable to the protectionist perspective. However, the mass starvation of animals in severe winters has provoked political and public unrest (ICMO 2006).

In winter, the limited availability of food and bad weather conditions undermine the health and welfare of the animals. This has led in severe winters to the death of many animals. Several times attempts have been made by animal welfare groups to change the hands-off policy and to prevent starvation by supplementing extra feed to the animals. In 1999, farmers even threatened to drop hay from aeroplanes. Every five years or so, politicians in parliament try to stop the starvation of the animals. In 2011, the Dutch Minister of Agriculture allowed them to be fed. Some political parties, for example the Liberal Party, would like to end the 'experiment' and to take all the animals out of the area to protect biodiversity. State Secretary Bleeker said that the 'experiment' had failed.

According to the rangers, intervention to prevent starvation will ultimately undermine the self-sufficiency of the population. They argue that high mass starvations also occasionally happen in other natural reserves, for example the Serengeti reserve in Africa (Vera 2009). An international advisory commission (ICOM 2006), however, advised the National Forest Service to intervene on the grounds of animal welfare, which, finally, ended its hands-off policy. The rangers now shoot an animal if its behaviour indicates that it will die soon. At the end of the winter of 2012/13, the Service shot 112 Heck cattle, 235 Konik horses and 1,136 red deer; and a further 33 Heck cattle, 48 Konik horses and 120 red deer also died of natural causes. This means that in April 2013 there were 3,086 animals left in the Oostvaardersplassen: 165 Heck cattle, 827 Konik horses and 2,094 red deer (Staatsbosbeheer 2012).

The shift from public debate to public communication

More than ten years after the introduction of the cattle and horses in 1983, as already indicated, numbers were large enough to attract attention when animals suffered during the harsh winter of 1995–96. The general public, farmers, park visitors and animal protectionists demanded supplementary feeding. In their view, according to the Animal Health and Welfare Act (Gezondheids- en Welzijnswet voor Dieren, 1992), the rangers had a responsibility to care for individual animals in all circumstances. Another argument against the management regime in the Oostvaardersplassen was that the welfare and health of the introduced animals was not solely determined by genetics, but also by the animals' experiences (Klaver et al. 2002). For years, the animals had depended on human assistance, and the sudden absence of this care was considered to be very stressful. Furthermore, opponents claimed that the situation in large reserves is far from natural: animals are fenced in and hindered in their migration, and there are no natural predators to 'take care of' sick animals (ibid.). Indeed, the rangers allow no medical intervention and there is no permanent veterinary care. In an interview, Jan Griekspoor, a fauna manager in the Oostvaardersplassen, admitted that there were indeed a number of initial problems with animals which were originally drawn from nature parks and zoos. According to Griekspoor, this was just a start-up problem; nowadays, they do as well as their naturally wild counterparts, the roe and the red deer. He claims that there is no need for supplementary feeding and medical intervention, because this could prevent the emergence of those qualities most suitable to the climate, the terrain and the food supply (ibid.).

What could explain the massive dissatisfaction with the fauna management in the Oostvaardersplassen? Reserve managers have suggested that, because Heck cattle look like cows and Konik horses look like riding-school horses, people think that they should be treated like these domesticated animals (ibid.). The general public compared animal welfare against an agricultural benchmark. Not every deviation from the riding-school model or the cow stable implies neglect. With unbrushed and matted coats, they seem to be neglected. The rangers themselves prefer to differentiate as little as possible between the originally domesticated animals (the Heck cattle and Konik horses) and the wild animals in their areas (the red deer and the roe). Vera is rather optimistic about the general public, and claims that a learning experience and proper education will contribute to a better understanding of de-domestication (Vera 2009).

In 2005, the public debate reached a climax when the Society for the Protection of Animals (SPA), an animal welfare NGO, sued the National Forest Service. In court, the SPA compared the deteriorating condition of cattle and horses in winter with the condition of farm livestock. They thus redefined the welfare and health conditions of cattle and horses in nature reserves in terms of experiences with domesticated animals on farms. The judge decided otherwise, and the SPA appealed and lost again. This verdict alienated many people from the government's switch in nature policy.

Case II: the Millingerwaard

The naturalness of the original area

Over the centuries, the Netherlands has had to protect its land from floods because it lies below sea level or below the high-water level of the major rivers. The dikes system determines the Dutch landscape to the present day. In the future, because of climate change, the water level is expected to rise. The traditional strategy of reinforcing the dikes will not work anymore, because it causes the peaty soil to dry out and the land to subside. For these reasons, the Dutch government decided on a new policy of flood risk reduction with regard to river management (Van der Brugge et al. 2005). This policy aims at creating more space for rivers and lowering high water levels, by means of deepening the forelands of rivers, displacing dikes further inland, lowering groynes in rivers, and enlarging summer beds. In the new policy, ecological restoration plays a key role. Many former agricultural areas are transformed into wetland reserves.

One of the first projects to turn former agricultural areas into wetland reserves was the Millingerwaard, situated east of Nijmegen near to where the river Rhine splits into the Waal and the IJssel. It is an area of 700 ha. (including 400 ha. of grazing land) on the south border of the Waal. In the past, this river meandered a lot and deposited sand on the river banks. Later, when the river changed its course, this sand was washed away. Nowadays, artificial dams keep the river in place. The west wind blows the sand, thus producing huge dunes. Until 1989, the southern bank of the river was an agricultural area. There were river meadows but also corn fields on which farmers used a lot of additives, like manure and pesticides. In the 1980s, the agricultural area diminished because of the removal of clay on an industrial scale. More and more farmers sold their land to the World Wildlife Fund and left the area. The minimal industrial activity also slowly disappeared.

How natural was the situation in 1989 at the beginning of the development of the Millingerwaard? The original area was relatively unnatural because it was part of a larger agricultural area. The river was tamed, and the sand from its banks as well as the clay, were used for industrial purposes. The reasons for changing the area into a nature reserve were technological: to protect the land from flooding and to use the rich clay reserves for industrial purposes.

Spontaneous nature

In the Millingerwaard, developers claim that its character is one of spontaneous nature. The land left bare after the removal of the clay, and the farmers' abandoned fields, have been taken over by plants and animals. Because of the work undertaken by farmers, the top-soil on their land had been enriched for centuries by the use of manure and other additives. This prevented the establishment of a stable ecosystem. Technological measures were taken to create a new natural infrastructure. The top-soil and trees were removed, and water holes and creeks were dug, but cycling routes and a campsite were also created. All kinds of information signs were erected. These technical interventions clearly go against the claim that the area has a spontaneous natural character, though there is some truth in it. Slowly, more and more sand-dunes appeared. After the first ten years, more than 400 species of plants were recorded in the Millingerwaard. Animals from other areas visited it and sometimes decided to stay.

Authentically wild animals

The open structure of the Millingerwaard vanished gradually because there were no grazing animals to keep it open. One hundred and twenty Galloway cattle, 160 Konik horses and some beavers were introduced into the area. To keep the animals within the rather small nature reserve, a fence was erected around the area. In the Millingerwaard, the image of an authentically wild animal functions as a model or icon for the Konik horses. In accordance with the ideal of the tarpan, herd managers Johan Bekhuis and Renée Meisner have tried to keep their herd of Konik horses in conditions that are as authentic or genuine as possible (Klaver et al. 2002). For example, they claim that cart-horse genes do not belong in a tarpan look-alike, and they try to deselect these genes in back-breeding experiments. In the rather small nature reserve of the Millingerwaard, this cannot be left entirely to natural selection. One is forced to mimic nature: ' "we select with human hands," states Meisner, "by looking with the eyes of a wolf" ' (ibid., p. 17). In this selection process, it is impossible to avoid errors and misjudgements entirely, and it remains a human artefact in which the 'naturalness' of the selection being made requires constant adjustment and reassessment. Basing the dominant ideal of genetic authenticity upon historical grounds causes problems. Striving for genuine, untouched, primeval nature leads to virtually impossible decisions. The question faced from the ecological restoration perspective is which historical point of reference one should take. Should we go back to the Middle Ages or go further back to the last interglacial epoch?

The shift from management to monitoring

Although the National Forest Service soon became the legal owner of the Millingerwaard, the Ark Foundation was responsible for fauna management. Recently FREE Nature, the Foundation for Restoring European Ecosystems, has become the new fauna manager. The foundation focuses on natural processes, like wind, water and grazing animals, and strives to develop self-regulating ecosystems in interlinked areas. Its ultimate goal is gradually to diminish human influence on the area, to replace management with monitoring. Grazing animals are the cornerstone of its management, and it has developed extensive ethical guidelines. Species-specific grazing affects the terrain – their presence must prevent the area from becoming choked by vegetation, specifically forestation, and therefore promote the variation that certain plants and animals need to establish themselves. The Ark Foundation has developed a strategy of selective breeding, in which humans function as predators. An important part of its herd management is to remove unfit animals and to replace them with suitable new animals, though the foundation has also learned not to destroy the social structure of the herd. For example, at the end of winter, the animals escape the flooding river by moving to higher ground. This only works when the experienced older mares can lead the herd.

In contrast to the Oostvaardersplassen, the Millingerwaard is a recreation area and tourists are more than welcome. Each year more and more people visit the nature reserve. This is an important part of the communication policy of the Ark Foundation, which deliberately established no official hiking trails. Everybody is free to roam the area in all directions. In my view, this has led to the establishment of an entertainment area in which the public could mistake a nature reserve for a zoo-like situation.

The shift from public debate to public communication

The development of new nature sometimes runs into problems because of lack of support among local residents (Buijs 2009). They experience the river landscape in a fundamentally different way than many governmental agencies and conservation groups. They regret the loss of many old (agri)cultural landscapes that disappear in the process of giving room to the river (Van der Heijden 2005) and feel that their sense of belonging to the landscape has suffered from the restoration practices (Drenthen 2008). In public debate, local residents often blame 'nature builders' for failing to recognize people's complaints. For example, some local residents near the Millingerwaard regularly cycle through this nature reserve to observe their favourite animals. If these animals have health or welfare problems, the residents immediately inform the fauna manager. He is officially not allowed to ask for the help of a vet but often gives in to avoid media attention.

In 2003, a survey was conducted into changing public perceptions and appreciations of river landscapes in response to ecological reconstruction projects in the floodplains (Buijs and van der Molen 2004; also see Buijs 2009). Local residents, holiday makers and people who did not have a specific relationship with a particular river landscape were interviewed. They were asked how the recent changes had affected their appreciation of the landscape. All groups said that they considered the new areas to be an improvement, that the new wetlands were visually more attractive than the old water meadows, and that they appreciated the increased accessibility of the areas. However, some local residents considered the traditional landscape just as natural as the new one, and most of them said that their feeling of place attachment had clearly diminished.

The behaviour of local residents over the years may be summarized by a number of simple rules: if animals are hungry, feed them; if animals are hurt, help them and call a vet; if animals die, remove their bodies; if animals are aggressive towards humans, kill these animals; make the area bike and hike friendly; it must be possible to touch and pet the animals; animals are not allowed outside the gate of the nature reserve. I have summarized these rules to show how far the Millingerwaard has digressed from a restored ecosystem.

Conclusion

Over the past 30 years, ecological engineers and rangers have developed the Oostvaardersplassen and the Millingerwaard into more mature nature reserves. This has attracted much attention in the media and also in scientific, political and public arenas. From the perspective of the protection of existing (primeval) nature, the development of new nature has resulted in fake nature. My conclusion, however, is that these two cases represent new examples of the typically Dutch situation in which all landscapes are manmade. The nature of these areas is not the result of restoration of some kind of former nature but of the development of new nature – which is not a fake nature.

Although this new nature is not fake, I claim that its developers have been faking. They have tried to link this new nature to the prehistoric period of aurochs and tarpans. They have wrongly presented this new nature as the return of nature by means of ecological restoration. Did the ecological engineers and rangers live up to the principle of the restoration ecology approach? In both cases, the principle of the naturalness of the original area was fake. In the Oostvaardersplassen, the original area was unnatural because it consisted of the bottom of Lake IJssel without its accompanying wet ecosystem. In the Millingerwaard, the original area was unnatural because the river Waal was controlled by dams. If the river had been able to meander, this would have meant time and again the destruction of recently formed ecosystems by floods. In both cases, the principle of spontaneous nature was fake because of technical interventions and the introduction of plants and animals into the areas. The principle of authentic animals fares no better. In the Oostvaardersplassen, the 'artificial' Heck cattle were introduced, and, in the Millingerwaard, humans function as predators of the Konik horses. In neither case did a shift from management to monitoring occur, although the rangers in the Oostvaardersplassen have made relatively more progress. Also, no shift occurred from public debate to public communication, although the Millingerwaard fauna managers have made relatively more progress by giving in to certain demands of the local residents.

In contrast to ecological restoration, the developing new nature approach does not advocate the naturalness of the original areas or spontaneous nature. It does not have to use authentic animals, and it favours a well-balanced mix of management and monitoring. Public debates are considered to be important indicators of possible problems in public communication. In both cases, the faking of the ecological engineers and rangers has hidden developing new nature behind the mask of ecological restoration and therefore caused many misunderstandings and misguided public debates – although it must be said that these debates have also helped to raise friendliness towards animals in society. Local residents and farmers were invited to fake the restored nature. It seemed that the public preferred animals in nature to behave according to a set of unnatural rules. The developing new nature perspective offers a way out of this problem by trying to learn from the views of local residents and farmers.

References

Buijs, A.E. and D. van der Molen (2004) 'Beleving van Natuurontwikkeling in de Uiterwaarden', *Landschap*, 21(3), 147–57.

Buijs, A.E. (2009) 'Public Support for River Restoration: A Mixed-Method Study into Local Residents' Support for and Framing of River Management and Ecological Restoration in the Dutch Floodplains', *Journal of Environmental Management*, 90(8), 2680–9.

Drenthen, M. (2008) 'Ecological Restoration and Place Attachment: Emplacing Non-Places?', *Environmental Values* 18, 285–312.

Elliot, R. (1982) 'Faking Nature', *Inquiry*, 25(1), 81–93. All citations in this chapter are from the reprint of 'Faking Nature' in A. Light and R. Holmes (eds) (2002) *Environmental Ethics: An Anthology* (Malden, MA: Blackwell), pp. 381–9.

FREE Nature (2013) www.freenature.nl (accessed 6 September 2013).

Gamborg, C. et al. (2010) 'De-Domestication: Ethics at the Intersection of Landscape Restoration and Animal Welfare', *Environmental Values*, 19, 57–78.

Gremmen, B. and J. Keulartz (1996) 'Natuurontwikkeling ter Discussie', *Filosofie en Praktijk*, 17, 14–27.

ICMO (International Committee on the Management of large herbivores in the Oostvaardersplassen) (2006) *Reconciling Nature and Human Interest: Report of the International Committee on the Management of large herbivores in the Oostvaardersplassen (ICMO)* (The Hague/Wageningen: Staatsbosbeheer), www.staatsbosbeheer.nl (accessed 6 September 2013).

Katz, E. (1992) 'The Big Lie: Human Restoration of Nature', *Research in Philosophy and Technology*, 12, 93–107.

Klaver, I. et al. (2002) 'Born to Be Wild: A Pluralistic Ethics Concerning Introduced Large Herbivores', *Environmental Ethics*, 24(1), 3–21.

Koene, P. and B. Gremmen (2001) 'Genetics of Dedomestication in Large Herbivores', paper presented at the 35th ISAE Conference, Davis, California.

Koene, P. and B. Gremmen (2002) *Wildheid Gewogen: samenspel van ethologie en ethiek bij de-domesticatie van grote grazers* (Wageningen: Wageningen University).

Staatsbosbeheer (2012) *Annual Report*, www.staatsbosbeheer.nl (accessed 6 September 2013).

Staatsbosbeheer (2013) *Annual Report*, www.staatsbosbeheer.nl (accessed 6 September 2013).

Stichting Ark (2013) www.arknature.eu (accessed 6 September 2013).

Van der Brugge, R. et al. (2005) 'The Transition in Dutch Water Management', *Regional Environmental Change*, 5, 164–76.

Van der Heijden, H.A. (2005) 'Ecological Restoration, Environmentalism and the Dutch Politics of "New Nature" ', *Environmental Values*, 14, 427–46.

Van Eerden, M.R. et al. (1997) 'Moulting Greylag Geese *Anser Anser* Defoliating a Reed Marsh Phragmitis Australis: Seasonal Constraints Versus Long-Term Commensalism between Plants and Herbivores', quoted in M.R. Van Eerden, 'Patchwork, Patch Use, Habitat Exploitation and Carrying Capacity for Water Birds in Dutch Freshwater Wetlands', PhD thesis, University of Groningen, 241–64.

Vera, F.W.M. (1988) *De Oostvaardersplassen. Van spontane natuuruitbarsting tot gerichte natuurontwikkeling* (Haarlem: IVN/Grasduinen-Oberon).

Vera, F.W.M. (2009) 'Large-Scale Nature Development: The Oostvaardersplassen', *British Wildlife*, June.

Vulink, J.T. and M.R. Van Eerden (2001) 'Hydrological Conditions and Herbivory as Key Operators for Ecosystem Development in Dutch Artificial Wetlands', quoted in J.T. Vulink, 'Hungry Herds: Management of Temperate Lowland Wetlands by Grazing', PhD Thesis, University of Groningen, 293–324.

8
Biodiversity and the Value of Human Involvement

Markku Oksanen

Introduction

Resurrecting an extinct animal species and releasing it into the wild is intentional human activity that increases biodiversity – or does it? Over the millennia, humans have played an important role in shaping the natural world by means of selecting and harvesting biological resources, creating new varieties and, disputably, new species (such as the dog) and manipulating the physical environment of their own and other organisms. The resulted forms of diversity are often fully dependent on humans. Many researchers have, however, been very reluctant to count human-dependent diversity as an element of overall biodiversity (for example Angermeier 1994). Theoretically, this reluctance rests on the assumption that there is a divide between the genuinely *natural* world with *natural* forms of life and the world brought about, or essentially modified, by human beings. Accordingly, there exists two kinds of biological diversity: (1) the variety of life that has been generated from nature's own forces and the evolutionary process of speciation; (2) the variety of life that has come into existence in virtue of, or has benefited from, human activities and, moreover, whose existence depends on these activities. Therefore, as far as this separation holds and the human-dependent diversity does not count, the resurrection of extinct species lacks an important justificatory foundation: resurrected animal species are not a welcome addition to the shrinking diversity of animals.

A major task in environmental philosophy – particularly in the early days of its development in the 1970s and 1980s – has been to highlight the value of the part of nature that is not affected by humans at all or where human impact is minimal. Despite the historical disregard for the issue, the justification of the value of human-produced and human-dependent biodiversity is relevant to environmental ethical theory. This subcategory of overall biodiversity includes the variability of cultivars, domesticated animals, cultural landscapes and restored and novel ecosystems. This is of the utmost significance to conservation policies not merely because of its vastness (see Perring and Ellis 2013) but also because of its importance to human well-being. At the same time, however, these policies embrace such political choices that social controversies are inevitable.

In this chapter I consider the idea of species resurrection and reintroduction in the context of biodiversity preservation. I construct a preliminary normative framework within which crucial human decisions concerning biodiversity are made and within which the risks of some policies are pointed out. Thus I shed light on the complex nature of the duty to protect biodiversity. The outcome of the analysis is a case-sensitive idea of biodiversity protection consisting of both negative and positive (including reparative) duties. Contrary to some environmental ethicists, I claim that human-dependent biodiversity is valuable and that there are moral reasons for carrying on with those practices on which this part of biodiversity depends. In other words, from the perspective of biodiversity preservation, the novel ideas of de-extinction and rewilding should not be dismissed out of hand.

Two commitments of biodiversity policy

'Biodiversity' is a shorthand for 'biological diversity'. Ever since the term was coined in 1986, it has been used to encapsulate the variety of life in all its existing manifestations on Earth. More formally, it covers diversity within species, between species and of ecosystems or habitats. Assuming this characterization, the duty to protect biodiversity is by its content obvious at first glance: the more there are diverse kinds of organisms at different taxonomic levels, in different habitats and in complex relationships and processes, the better. There are, however, many controversies that lurk behind this intuition. These controversies stem partly from the lack of a shared operational definition of 'biodiversity' and the consequential failures to measure it (Sarkar 2002; Koricheva and Siipi 2004; MacLaurin and Sterelny 2008, pp. 132–48), partly due to the genuine disagreements over the comparative value of different units of it.

To make sense of the duty to protect biodiversity, the concept of 'biodiversity' must first be established. As I see it, the following two commitments are regularly made in defining 'biodiversity' and forming the policies of biodiversity preservation. A statement by the famous biologist Ernst Mayr offers a good starting point for the first commitment:

> The most impressive aspect of the living world is its diversity. No two individuals in sexually reproducing populations are the same, nor are any two populations, species, or higher taxa. Wherever one looks in nature, one finds uniqueness. (Mayr 1997, p. 124)

The diversity of nature can thus be understood in terms of the individuality of living beings. It is, however, clear that neither each and every sexually engendered individual nor each and every individual of a microscopic species deserves, for instance, special attention granted by a preservation policy. It is simply beyond human powers to preserve all the strange outcomes of the evolutionary process, some of them carrying a gene or chromosome mutation with no hope of survival. A rigid individualist approach to biodiversity preservation would require the maximization of the number of living individual beings. If all forms of life – sentient animals, plants and even micro-organisms – were included in such an approach, it would be unfeasible. Saying this I do not want to devalue micro-organisms but to emphasize that without qualifications the approach would demand humans to live lives of non-corporeal beings.

A distinctive feature in biodiversity policies is that the subjects of concern are general categories or types (populations, species, biotopes, ecosystems) rather than individuals.[1] Both for historical and for practical reasons species are regarded as the most important components of biodiversity, which is usually measured in terms of species richness (Gaston and Spicer 1998, p. 10). Typically biodiversity policies take the form of some individuals being favoured over others solely because of their species membership (even though the treatment of individuals as instances of species can be constrained by supplementary moral principles such as the principle of non-maleficence). From the non-anthropocentric individualist and egalitarian perspective this policy is, however, considered ethically disturbing (Regan 1988; Taylor 1986; Rawles 2004).

The second commitment regularly made in defining 'biodiversity' is that, although the duty to protect biodiversity implies that a certain state of affairs or condition is worth maintaining and protecting, it should not be understood so as to presuppose that there is a fixed number of species to be preserved. To think this would require the basis of a teleological view of the evolutionary process. The aim in biodiversity preservation is thus something different than bringing about such a state of affairs, the components of which can be thoroughly listed. The evolutionary process has taken place on Earth over a course of time and there are more different forms of life now than there was at the beginning of the process, despite the five previous mass extinctions. The process is, however, aimless, and there is no grand plan or final end dictating which forms of life prosper and which die off. Therefore, the dynamic aspects of duty should be stressed because the existence of biodiversity is linked to the idea that processes, not entities, are ultimately the objects of biodiversity policies, as extant biodiversity results from these processes. These policies aim to ensure the viability of the evolutionary process which requires that the genetic lineage in actual organisms from the present to the future is not deliberately broken and that further diversification and speciation can occur.

What characteristics make a site or a species relevant to biodiversity policies is to some extent a matter of political choice. Everything cannot be conserved or preserved and therefore entities, processes and phenomena have to be classified and ranked. This fact alone has triggered many conflicts regarding biodiversity policies. The focus on individuals, genes, species and ecosystems has not eliminated the emergence of these conflicts, and apparently it cannot do so. The divide between human-dependent and natural biodiversity is a minor strain within the larger framework. In summary, biodiversity is deeply and essentially a value-laden concept.

The nature of the duty to protect biodiversity

Given the two commitments above, that is the focus on types and the open-ended nature of evolutionary processes, we can start analysing the nature of the assumed duty to preserve biodiversity. John Passmore (1980, p. 125) proposes that 'the existing degree of diversity in an ecosystem ought not to be modified without careful consideration'. This proposal is not unambiguous. It is *conservative* in a sense that it favours 'the existing degree of diversity' but the issue whether it forbids humans both from reducing and from increasing biodiversity is left undeveloped. In this section, I shall distinguish three competing views as to how this duty can be understood: global negative duty, global positive duty and case-sensitive duty (which has both positive and negative elements).

Global negative duty

Virtually all environmental philosophers accept the view that humans ought to forbear from activities that decrease the variability of species, subspecies or ecosystems. This can be called 'global negative duty' (GND) and it forbids the active destruction of biodiversity. The duty is global because it is independent of the site or the locality; it is the abstract idea – for example, the number of species – that matters. In its general form, this obligation is obvious and does not give rise to any serious disagreement. It is apparent that if hunting of the brown bear reduces the size of the bear population or brings it to the verge of extinction, then hunting is not morally right. Likewise, if cutting down an old-growth forest would risk the future of the rare species populations living there, then the forest should not be cut down. However, when humans use some unit of biodiversity, the point must be determined as to when this use becomes susceptible and reduces biodiversity in the morally relevant sense – a matter that tends to stir continual controversies in the management of natural resources.

In general, the duty to protect biodiversity in this sense is a prima facie duty. This means that humans may decrease the extant biodiversity for good reasons, for example eradicating alien species. A more general set of reasons for biodiversity reduction relates to meeting human needs. This qualification of GND is commonly accepted by environmental ethicists, such as Arne Naess (1989, p. 29), Paul Taylor (1986) and Bryan Norton, whose stipulation encapsulates the general mentality: 'species should be saved provided the social costs are acceptable' (Norton 1987, p. 237). Of course, the social costs proviso can be interpreted in many ways and it requires further specifications, such as what kind of risk has been taken when a species is allowed to become extinct and how the benefits, burdens and risks are to be distributed among the stakeholders, in particular among humans.

To meet these qualifications requires that organisms and species are ranked into deontic hierarchies: roughly, those that are worth protecting, those that must be wiped out and those that are insignificant. This leads to a situation that could be called the 'smallpox virus dilemma'. There are obnoxious species most humans want to get rid of for good, such as the smallpox virus, the pubic louse or the parasitic trematoda;

but to sanction extinction of one species may result in a slippery slope: letting one species be eradicated allows that any single species could in principle be eradicated. This dilemma can be used, however, either as an argument for the total preservation of all extant species or as an argument against any single preservation plan when the preservationist admits the possibility of the disvalue of species. The slippery slope argument is unintuitive because it is clear that both policies, wiping out any species and preserving every species, are extremely costly for humans.

Despite these reservations concerning its limits, I would like to call GND *the basic norm* of biodiversity preservation: it is the foundation that ought not to be violated except under extraordinary circumstances. It commands us to perform those actions, to implement those policies and to establish institutions that are the most favourable in regard to the efforts to protect the extant biodiversity. Those actions, policies and institutions that have a decreasing effect on biodiversity violate this duty and therefore require a proper justification. On general terms it is rather easy to characterize the basic norm. It is more difficult to determine what it prescribes in real-life situations and how, for example, people can learn to manage the natural environment without compromising this duty. When it comes to the resurrection of extinct species, GND in itself does not tell us whether resurrection is good or bad. However, if humans decide to revive species, it ought not to impoverish nature's diversity.

Global positive duty

For many philosophers, the concept of value is linked to an obligation to maximize value, that is, to increase the number of valuable things in the world as much as possible. This raises a question: is it desirable that humans aim to increase biodiversity? If the answer is positive, we have the global positive duty (GPD) at hand. This requires aiming at the maximization of biodiversity. This is, however, an interpretation that many conservation biologists and environmental ethicists are willing to reject outright (for example, Ehrenfeld 2006).

In the light of the debate on genetic engineering, it appears that many conservation biologists and environmental ethicists are extremely dubious about the idea that humans may use their intellectual and technical capabilities for maximizing biodiversity without any constraints whatsoever. The principal constraint is that the creation of new forms of biodiversity must not result in a reduction of the already existing diversity. Some might go even further: despite the presumption that biodiversity is a positive value, there is an obligation not to increase biodiversity.

There are several rationales as to how to justify the restraint of not increasing biodiversity, including species resurrection. It is customary to refer to the *method* of increasing diversity, for example, whether it occurs by means of modern biotechnology and genetic engineering, and to claim that wrongness is essentially related to the mode of technology (for example, Lee 2004, pp. 154–6). Similarly, the resurrected species is less valuable because of its unnatural origin. Another rationale for banning the production and introduction of new forms of life in the wild points to the *consequences* to biodiversity: there is a risk of homogenization so that the new species might displace native ones and diminish the overall biodiversity (for example, Angemeier 1994, p. 602). Again, the resurrected and introduced species will most likely affect the extant flora and fauna. Perhaps the most disapproving stance is the one that regards the production and introduction of new species as an intrusion upon the integrity of natural ecosystems (Westra 1994). To transfer the polar bear into the Antarctic ecosystem, for instance, is a clearly suspect way of increasing biodiversity there, although some argue for the so-called 'assisted migration' (or 'managed relocation') as a way to adapt to global warming (Minteer 2012, pp. 166–70; Sandler 2010). Accordingly, the resurrected species might be seen to harm the ecosystem if the species has not dwelled in the place ever before, or for a long period of time, so that it cannot be considered as an integral part of the functioning of that ecosystem.

It has been an essential part of human cultural evolution that humans have caused both decrease and increase in biodiversity. If GPD is denied categorically, it follows that people must act on the principle of letting nature take its course and that the outcomes would be rather unfortunate for humans. Let us consider two examples of anthropogenic increase in biodiversity: landraces and cultural landscapes.

The so-called 'landraces' of plants and animals are examples of human-dependent diversity of life. For example, the International Rice Genebank stores 117,000 varieties of cultivated rice (IRRI 2013). These varieties mainly result from 'a continuous recombination process between two independently domesticated subspecies [of Asian rice, *Oryza sativa*], so-called *indica* and *japonica*' (Bellon et al. 1997, p. 280). It is probable that the farming and breeding of rice has increased rather than decreased its genetic variability, and that there are more varieties now than there would have been if humans had not functioned as a selective force (on the problems of this claim, see Brush 2004, pp. 64–6). In biodiversity preservation, the attention that the genetic variability within domesticated species requires is rather exceptional. Paying most attention to developing high-yielding, genetically uniform varieties easily leads to negligence towards less productive varieties and so narrows the genetic make-up of the species (Kloppenburg and Kleinman 1988, p. 6; Perlman and Adelson 1997, pp. 58–60). Similarly, any attempt to protect the genetic variability of rice by means of forbidding the use of less productive varieties is plainly ineffective. It would rather mean letting these varieties vanish – a kind of indifference to a decrease in variability. So, it seems necessary that the use of rice varieties continues despite the paradoxical nature of this option: farmers have actively selected the specimens that are the most suitable for their needs and hope they will reproduce. In practice, human involvement has resulted in great variability (Bellon et al. 1997, p. 272; Darwin 1998 [1859], p. 377).

As a second example, consider changes in the decreasing biota of the countryside in Europe. Vuorisalo and Laihonen (2000, p. 192) have pointed out that 'in Finland, habitat change has replaced hunting and persecution as the main threat to biodiversity'. As to the countryside, traditional landscapes and cultural habitats are disappearing. At the end of the nineteenth century two-thirds of agricultural land was used for grazing animals, which was classified as meadow. During the following one hundred years, grazing ended and meadows were converted into fields, or they were afforested. According to the 2010 Finnish National Survey, for 22.3 per cent of endangered species the primary threat is the overgrowth of open habitats (Rassi et al. 2010). Thesebiotopes which are by Finnish standards rich survive largely by virtue of the substitutive action taken by conservationists.

The varieties of rice and traditional landscapes are examples of human-dependent diversity of life; they have more or less come to exist by means of human selective and reproductive activities without which they are likely to go extinct or disappear. Although the genus *oryza* exists in wild populations, human selection and management has apparently produced great variability within a single species. Given this, the disappearance of cultivars can be defined as human-caused, even though it can be either unintentional or intentional. When it is intentional, it is within human power to influence the existence or disappearance of a variety, and humans know about this possibility: when these conditions hold, they are responsible for these outcomes in terms of omissions and actions. This intentional disappearance can happen either when people let these forms of life vanish by refraining from performing certain acts that they know are needed to maintain diversity, or when people simply wipe varieties out by performing certain acts known to have such an effect. In summary, there is a distinction between letting a species go extinct (Chessa 2005) and making it extinct (see Shapiro 2001, pp. 86ff.). It is not clear, however, whether GPD allows for reversing extinction or not. Some of the main reasons were mentioned earlier in this section and they are related to the nature of the method, the consequences for the extant biodiversity and the idea of integrity. Also the question about the fundamental nature of resurrected species is an open issue: if the emphasis is on species, resurrected animals should then be classified as species in their own right so as to qualify as an addition to the overall biodiversity. If the emphasis is on the intraspecific diversity, then there is a possibility of their being additions to biodiversity. However, it is not clear to which species a hybrid of woolly mammoth and elephant belong.

Case-sensitive duty

On the basis of the above analysis, global approaches run into problems. In nearly all cases the anthropogenic decrease in biodiversity is unacceptable; in some cases the anthropogenic increase of biodiversity is acceptable. Thus the approach to biodiversity conservation should be case sensitive in an overt manner. Contrary to those theorists who claim that human-dependent biodiversity is of less value, if not of disvalue, I argue that an adequate theory of environmental ethics should appreciate human-dependent biodiversity and assign it a proper place in moral deliberation. Still, it seems reasonable to assume that this deliberation should be constrained by the duty of being cautious of, or of fully abstaining from, producing and introducing biological novelties because of uncertainties and the unpredictability of consequences for the extant biodiversity. Combining these two considerations leads to the conception that the duty to protect biodiversity should contain *a permission* to increase biodiversity and maintain the biodiversity that humans have produced in the course of history.

Because there seems to be no unambiguous judgement of this principle of protecting biodiversity, the focus is on cases. The case-sensitive approach (CSA) addresses situations in which humans have to decide how to affect biodiversity.

> CSA: If the result of doing P is that there is *less/more* biodiversity on Earth than in the case of not doing P, it can be either wrong or right to do P, depending on other relevant factors.

This formulation is not very helpful in providing us with a precise guideline; nevertheless, it illustrates the complexities in biodiversity policies. What are the relevant factors in this formulation? Often the reduction in biodiversity in some location takes place on the basis of the assumption that to do so is to protect the original diversity and thus the overall diversity that had prevailed in earlier times. The problem encountered here is the *problem of exotics*. A widely held view among ecologists is that the human-caused increase in biodiversity at the assemblage level is undesirable because alien species are one of the leading causes of the extinction of endemics at the global level (Committee on Noneconomic and Economic Value of Biodiversity 1999, p. 28). For example, Wilson contends that the worst anthropogenic change in an ecosystem was that which resulted in the extinction of many species in Lake Victoria. It was caused by the British colonists who introduced a new fish for sport, the Nile perch (Wilson 1994, p. 244). Aliens can out-compete the authentic populations, but they can also be relatively harmless in this respect. Most of the basic food plants have spread all over the world; so, it is not simply the case that relocating species is categorically damaging to humans or to ecosystems. Thus humans face the question of whether to eradicate the exotic species or to welcome it as an addition to the original pool of species. Perhaps it is apposite to characterize the duty to protect biodiversity also as the duty to recognize all the dimensions of the ways in which human behaviour affects nature.

The basic concern about formulating CSA as above is that it is a sort of tautology and so is unable to tell us what exactly should be done in regard to biological diversity. *But this is probably the best we can get when we aim at protecting virtually everything that exists on Earth.* In defining the nature of the duty to protect biodiversity there are many factors to be taken into account. In Table 8.1, I illustrate the relationship between the nature of human involvement and its effect on biodiversity. Above I considered two cases in which a passive hands-off, or non-interventionist, policy would probably have a decreasing effect on the extant biodiversity. In regard to this issue the policy recommendation, providing the desire to protect biodiversity, differs from the case in which active involvement, in the form of using biodiversity resources, is the main driver of the biodiversity loss. Traditionally, this has required abstaining from using the endangered species, that is, humans should be consciously passive in regard to the protected species. Whether this kind of non-interventionist policy is invariably the best way to ensure the future existence of biodiversity has been questioned. At least in the case in which biodiversity depends on human involvement, it is not

Table 8.1 The relationship between human involvement and biodiversity

Nature of human involvement	Effect on biodiversity	
	Increase in biodiversity	**Decrease in biodiversity**
Passive	• Manipulation of natural evolutionary processes (for example isolation) • Traditional preservationism and species protection legislation • Non-interference • Indifference toward bioinvasions	• Neglect of the maintenance of high-diversity human landscape • Natural local/global extinctions • Indifference toward bioinvasions • Indifference toward changes in the physical environment
Active	• Species and habitat conservation • Creation of new species and varieties • Introduction of exotics • Assisted migration • For other cases, see pp. 161–4	• Unsustainable hunting and fishing • Over-harvest of natural plants • Dramatic changes in land use (habitat fragmentation, deforestation, desertification, draining of swamps) • Introduction of exotics

the best policy. Whether the same applies to natural biodiversity, I leave aside here.

In summary, the idea of the biodiversity preservation promoted here emphasizes the need to consider the cases individually and the importance of taking into account all relevant factors. Having said this, it appears that the duty to protect biodiversity is not a purely negative one that requires nothing but restraining oneself from action that depletes or actively destroys biodiversity. The duty is something that falls between these options and thus it lacks general, substantial content (cf. Michael 2001, p. 154). This aspect is indicated by the fact that my analysis ended up in a kind of tautology according to which the nature of the duty depends on other relevant factors that cannot be universally and a priori characterized. In essence, biodiversity preservation relates to the pluralistic conception of what is valuable. As often in pluralism, it is rather difficult to provide clear and unambiguous rules of appropriate conduct. Therefore, any attempt to define the nature of the duty to protect biodiversity in a more substantive way goes astray. Decisions concerning biodiversity necessarily involve trade-offs of whatever kind of diversity we desire to exist and, therefore, we should focus on both cases and case-types instead of focusing on this duty at a highly abstract level.

Ways of biodiversity increase

To some extent, though it varies in time and place, biodiversity is human-dependent. Although humanity is the main driver of the current biodiversity loss, there are at least six ways in which humans can increase biodiversity.

(a) Intentional inaction to allow evolutionary speciation to continue

Because of human presence, power and technical advancement, the standard method in biodiversity preservation is the hands-off policy, or letting things be as they are. Intentional inaction is actually a chosen action to give up those practices that have brought some populations to the brink of collapse. Intentional inaction can thus allow evolutionary speciation to continue and existing species to survive, though in some cases inaction might result in biodiversity loss.

(b) Creating brand new forms of life and biotopes

Synthetic biology is 'the design and fabrication of biological components and systems that do not exist in the natural world' (Synthetic Biology 2013). Only if it is assumed that the duty to protect biodiversity is global and positive in kind, the creation of brand new forms of life can be thought of as morally right or even praiseworthy. However, talk about protecting biodiversity should then be complemented, if not replaced, with speaking about inventing or creating biodiversity, as it would depict the project to protect biodiversity better. As a general rule, it appears fully acceptable to go on to develop new varieties from already domesticated species. However, there are some rightly imposed legal and moral restrictions concerning method (the regulation of genetic engineering), outcomes (not to create an animal in permanent pain) and use (its use in natural settings is subject to licence). Instead, developing entirely new species or subspecies, something that could be called *de novo* species that are meant to live in the wild and to compete with the wild species, is an issue to which most environmentalists would take a disapproving stance, for the reasons mentioned earlier.

If it is assumed, as I do in this chapter, that traditional human-dependent biodiversity is worthy of protection, the question is raised as to why *novel* human-created species or subspecies lack the value that *natural* species have, or why novelties can even be regarded as having no value. What is the morally relevant difference between these two cases? The defenders of genetic engineering typically claim that there is no morally relevant difference and that new varieties can be made. The opponents of genetic engineering deny this assertion. In my view, humans should be cautious about making changes that radically differ from what has existed for a long time, because of the limited capability to foresee the consequences to the full. The other issue is that the veiled objective of producing genetically modified organisms is to produce organisms with stable and controlled qualities. In other words, it results in homogeneity within species and is therefore contrary to the continuous contingency of natural reproduction. It is humanly imposed essentialism: each organism has its ideal type conceived and shaped by humans, whereas natural selection and conventional domestication results in heterogeneous populations and biological individuals (Mayr 2002, p. 91).

(c) Re-creating extinct species, subspecies (variations) and biotopes

As stressed in the Introduction to this book, the idea of re-creating extinct forms of life fascinates people. At the same time, many environmentalists are worried about it, mainly because it tests the conventional wisdom of conservation biology about the irreversibility of extinction. There are two main methods of realizing resurrection: back-breeding and gene-technological methods. Back-breeding aims to reproduce vanished varieties or species by conventional breeding. Gene technology attempts to generate a clone of the original source from a DNA sample of an extinct species. Provided that the resulting organisms are members of some species, and not ontological anomalies, this method might mitigate the situation of endangered species and revitalize the extinct ones. For some biologists, the resurrection of extinct species, especially vertebrates that might adopt a key role in an ecosystem and thus contribute to rewilding it, is a worthy goal (Zimov 2005; Donlan et al. 2006; Svenning 2007).

Others oppose this for various reasons. One important objection rises from the fear that it reduces the significance of other means of preservation (Ehrenfeld 2006, pp. 730–1). This fear is justified only if the resources and funding needed for the development the new technique decrease the traditional investments in research and conservation. Briefly, the method is complementary and the last resort in 'preserving' the species. It is virtually impossible to obtain a DNA sample of each extinct or endangered species and save them in a gene bank for decades and then to reintroduce the species in the wild. Providing that the scientific and technical problems are solvable, whatever other problems remain they are likely to be similar to those of reintroducing locally extinct species, including that of re-establishing a population of recreated species. Gunn (1991, pp. 298–301) has, however, questioned the idea that the re-created individual is a genuine member of the original population: the ecological position of the re-created species is different from its historical precedents. Thus, a slightly modified but still transgenic organism should not be conceived of as a specimen of an original species. As I see it, even though the re-created species was not of the same value as the 'authentic' one, it does not detract from the value of this enterprise. In practice, this point reminds me of objections to zoos as safe havens for animal species. According to the critics, however, in a few generations zoo animals develop into a breed of their own and differ from the populations of the same species in the wild (Lee 2005).

(d) Reintroducing a locally disappeared species

The issue of reintroduction of species into ecosystems from which they have disappeared is primarily a local matter. There is a growing literature of ecological restoration that addresses the ecological preconditions of reintroduction of a species and related forms of action, such as eradicating competing non-native species. In practice, the restoration and the reintroduction of an extinct species has turned out to be a hot topic culturally, socially and politically. Case studies on the wolf (Sharpe et al. 2001), the beaver (Gamborg and Sandøe 2004) and prairie restoration in Chicago (Gobster and Hull 2001), among others, have been carried out, which show that sometimes the authentic species and landscapes are welcome and at other times they are not. Unlike the integrity argument in case (b) to oppose genetic engineering, in cases (c) and (d) it is the same technological achievements that could be used to restore the lost integrity.

(e) Maintaining extant human-dependent species, subspecies (variations) and biotopes

The active maintenance of human-dependent biodiversity refers to, in particular, traditional management systems that produced plenty of diversity. Without human activity, these human-modified habitats and cultivars will vanish.

(f) Preventing natural extinction of populations or species

Natural (not-human-influenced) extinctions are divided into two categories in terms of causes: changes in the environment and changes in the interactions between species (Johnson 2001). Some scholars have argued that preservationism ignores the time dimension and ultimately attempts to halt incessant change and to freeze the present state of affairs (see Callicott 1998, p. 349). One method to achieve this is to stop those disappearances of species that have no involvement with humans. In my view, the interference with the natural world to prevent natural extinction is acceptable, mainly for two reasons. First, it is reasonable to presume that humans play a role in the current loss of biodiversity. Namely, the prevalent trend nowadays is that the majority of species and population extinctions are caused by humans and it is virtually impossible to detect other possible causes of extinction. As a general policy, therefore, it is better to prevent species from dying out, as far as it falls within human capabilities. Second, when humans aim to maintain the existence of a species, that species is not intentionally planned and created; in this respect, case (f) differs from case (b) and is a far less serious intervention in the processes of nature.

* * *

Some of these cases are clearly global, others are local, and the third kind have both of these features. Some cases might be of more use as material for science fiction, but the rise of a more intervention-oriented approach to conservation (Hobbs et al. 2011), such as the development of genetic engineering and ecological restoration, will probably open the way to new problems. It is again apparent that cases whose realization falls within human capacity are morally intricate and socially controversial. So I think the best approach is provided by presuming an ambiguity toward the duty to protect biodiversity, rather than by espousing either global positive duty or global negative duty.

Concluding remarks

In this chapter I have analysed the nature of the duty to protect biodiversity. I have claimed that this duty is not merely negative, but also positive, and requires the protection of human-dependent biodiversity. I have not discussed the value of biodiversity. In general, I think the idea that biodiversity has intrinsic or immediate value – in the same sense as, say, humans do – is untenable; complexities in nature are of instrumental value, as they are also for both humans and other living beings. Despite this axiological position it is not unintuitive to claim that total biodiversity is something that should be taken into account. Accordingly, I find it reasonable that the duty to preserve biodiversity is neither a purely negative duty not to destroy existing species, nor a positive duty to increase biodiversity wantonly; instead, this duty is something that falls between the extreme options. Given that human-produced biodiversity has characteristics of artificiality, its destruction is not routinely approved. Whenever biodiversity is reduced, this action must be justified because it violates the grounding norm of biodiversity protection. Furthermore, whenever biodiversity is intentionally increased, attention must be paid to three following features: the history of locality, the methods and the consequences.

Authenticity as a property of the object. Species can be characterized as being authentic, or native, to a place (Norton 2001, p. 223). This is a spatio-temporal concept denoting the fact as to whether a species has ever occupied a place without human intervention. Authenticity is a relevant aspect of biodiversity policies because it delimits the scope of human choice so that there should be continuity between the past and the present with respect to a particular place. The authenticity, however, should be qualified with a condition of benignity. Some species are harmful in a sense that their presence in an ecosystem reduces its diversity. When thinking about increasing *local* diversity, this possibility calls for special attention. Those species that are native or authentic to a place take precedence within preservation. Likewise, efforts must be made to eradicate those exotic species that constitute a threat to the array of authentic species.

The motive of the agent. In general, there is a good reason to be made that humans should repair the damage they have caused to biodiversity. Therefore, if they have the capability to re-create extinct species, this possibility may have to be realized and thus rectify the harm done. It is notable, however, that this criterion is applicable only to cases in which humans have played a crucial role. Accordingly, this does not oblige humans to re-create the dinosaur, though it can oblige them to re-create the dodo and the wild horse. The basic problem here is that the cause of the extinction is not obvious. Palaeobiologists are intensively studying how the greatest mammals became extinct over the past 100,000 years: different animal species and different populations seem to require different explanations (Barnosky et al. 2004). The fate of the woolly mammoth is an example of extinction from several causes such as climate and habitat change and human predation (MacDonald et al. 2012). Others consider the extinction of the woolly mammoth to be unsolvable while recognizing that humans have contributed to the demise of the wild horse, steppe bison and many marsupial species (Lorenzen et al. 2011; McGlone 2012). In the context of motivation, the moral issue about species re-creation turns into an empirical issue about the drivers of historical extinctions and human culpability.

The action of the agent. The means of increasing biodiversity can be either direct or indirect and whichever method we choose may also indicate human attitudes to living beings. In the case of genetic diversity, it seems useful to distinguish between protecting the natural processes and the social processes that both have a propensity to increase intraspecific biodiversity. Varieties that result from natural processes are ingredients of natural evolution. Social processes are usually conceived of as being something different from natural processes. Often we assess how close social processes are to natural ones, and thus some methods of producing diversity are conceived of as being more natural than others. Traditional plant and animal breeding is a case in point: we enhance the qualities we find useful. Is it, however, possible to keep those human processes that maintain and produce biodiversity in the traditional ways separate from those acts of creating biodiversity by using modern scientific means, for example genetic engineering? If we think that it is possible, we need to define the morally crucial difference between the organisms that result from traditional forms of biotechnology, including domestication and breeding, and those that are produced by means of modern biotechnology. I do not know whether it is possible to do this. But if there is a difference, it seems to lie in the origin of the variety and particularly in the technique used to bring about the variety.

Even though human-dependent biodiversity is of value, the duty to protect biodiversity should not be understood to give humans an open licence to increase biodiversity in any way they like. It is best to be conservative in both introducing new species and eradicating existing ones. The issue of resurrecting those species whose extinction is anthropogenic is, however, more problematic. My argument has been that re-created animals can be parts of the overall biodiversity. Nevertheless, biodiversity is a value-laden term and different kinds of biodiversity units are of different value. Thus it may well be that the case for human-supported animal species is weaker than is sometimes thought, and that restorative duties are not perfect.

Note

1. The emphasis on types rather than tokens extends to habitat conservation. Traditional wilderness preservation has concentrated on protecting unique,

unmodified places, whereas biodiversity policies have an interest also in places as types of biotope (Sarkar 1999, p. 406) or, as in biodiversity offsetting, ecosystem functions.

References

Angermeier, P.L. (1994) 'Does Biodiversity Include Artificial Biodiversity?', *Conservation Biology*, 8, 600–2.

Barnosky, A.D., et al. (2004) 'Assessing the Causes of Late Pleistocene Extinctions on the Continents', *Science*, 306, 70–5.

Bellon, M.R., et al. (1997) 'Genetic Conservation: The Role for the Rice Farmers' in N. Maxted et al. (eds) *Plant Genetic Conservation: The In Situ Approach* (London: Chapman & Hall).

Brush, S.B. (2004) *Farmer's Bounty: Locating Crop Diversity in the Contemporary World* (New Haven: Yale University Press).

Callicott, J.B. (1998) 'The Wilderness Idea Revisited' in J.B. Callicott and M.P. Nelson (eds) *The Great New Wilderness Debate* (Athens: University of Georgia Press).

Chessa, F. (2005) 'Endangered Species and the Right to Die', *Environmental Ethics*, 27, 23–41.

Committee on Noneconomic and Economic Value of Biodiversity, National Research Council (1999) *Perspectives on Biodiversity: Valuing Its Roles in an Everchanging World* (Washington, DC: National Academy Press).

Darwin, C. (1998 [1859]) *The Origin of Species* (Oxford: Oxford University Press).

Donlan, C.J., et al. (2006) 'Pleistocene Rewilding: An Optimistic Agenda for Twenty-First Century Conservation', *The American Naturalist*, 168, 660–81.

Ehrenfeld, D. (2006) 'Transgenics and Vertebrate Cloning as Tools for Species Conservation', *Conservation Biology*, 20, 723–32.

Gamborg, C. and P. Sandøe (2004) 'Beavers and Biodiversity: The Ethics of Ecological Restoration' in M. Oksanen and J. Pietarinen (eds) *Philosophy and Biodiversity* (Cambridge: Cambridge University Press).

Gaston, K.J. and J.I. Spicer (1998) *Biodiversity: An Introduction* (Oxford: Blackwell).

Gobster, P.H. and R.B. Hull (2001) *Restoring Nature: Perspectives from the Social Sciences and Humanities* (Washington, DC: Island Press).

Gunn, A.S. (1991) 'The Restoration of Species and Natural Environments', *Environmental Ethics*, 13, 298–301.

Hobbs, R.J. et al. (2011) 'Intervention Ecology: Applying Ecological Science in the Twenty-first Century', *BioScience*, 61, 442–50.

IRRI (International Rice Research Institute) (2013) http://irri.org/index.php?option=com_k2&view=item&id=9960:the-international-rice-genebank-conserving-rice&lang=en (accessed 11 June 2013).

Johnson, C.N. (2001) 'Natural Extinctions (Not Human Influenced)' in S.A. Levin (ed.) *Encyclopedia of Biodiversity* (San Diego: Academic Press).

Kloppenburg, J.R. Jr and D.L. Kleinman (1988) 'Plant Genetic Resources: The Common Bowl' in J.R. Kloppenburg Jr (ed.) *Seed and Sovereignty: The Use and Control of Plant Genetic Resources* (Durham: Duke University Press).

Koricheva, J. and H. Siipi (2004) 'The Phenomenon of Biodiversity' in M. Oksanen and J. Pietarinen (eds) *Philosophy and Biodiversity* (Cambridge: Cambridge University Press).

Lee, K. (2004) 'There is Biodiversity and Biodiversity: Implications for Environmental Philosophy' in M. Oksanen and J. Pietarinen (eds) *Philosophy and Biodiversity* (Cambridge: Cambridge University Press).

Lee, K. (2005) *Zoos: A Philosophical Tour* (Basingstoke: Palgrave Macmillan).

Lindgren, L. (2000) *Island Pastures* (Helsinki: Metsähallitus & Edita).

Lorenzen, E.D., et al. (2011) 'Species-Specific Responses of Late Quarternary Megafauna to Climate and Humans', *Nature*, 479, 359–65.

MacDonald, B.M., et al. (2012) 'Patterns of Extinction of the Woolly Mammoth in Beringia', *Nature Communications*, 3, 893.

MacLaurin, J. and K. Sterelny (2008) *What Is Biodiversity?* (Chicago: University of Chicago Press).

Mayr, E. (1997) *This is Biology: The Science of the Living World* (Cambridge, MA: Harvard University Press).

Mayr, E. (2002) *What Evolution Is* (London: Phoenix).

McGlone, M. (2012) 'The Hunters Did It', *Science*, 335, 1452–3.

Michael, M.A. (2001) 'How to Interfere with Nature', *Environmental Ethics*, 23, 135–54.

Minteer, B.A. (2012) *Refounding Environmental Ethics: Pragmatism, Principle, and Practice* (Philadelphia: Temple University Press).

Naess, A. (1989) *Ecology, Community, and Lifestyle: Outline of an Ecosophy*, ed. and trans. D. Rothenberg (Cambridge: Cambridge University Press).

Norton, B.G. (1987) *Why Preserve Natural Variety?* (Princeton: Princeton University Press).

Norton, B.G. (2001) 'What Do We Owe the Future? How Should We Decide?' in V.A. Sharpe et al. (eds) *Wolves and Human Communities: Biology, Politics, and Ethics* (Washington, DC: Island Press).

Oksanen, M. (1997) 'The Moral Value of Biodiversity', *Ambio*, 26, 541–5.

Oksanen, M. (2007) 'Species Extinction and Collective Responsibility' in Z. Davran and S. Voss (eds) *The Proceedings of the Twenty-First World Congress of Philosophy. Vol. 3. Human Rights* (Ankara: Philosophical Society of Turkey).

Passmore, J. (1980) *Man's Responsibility for Nature*, 2nd edn (London: Duckworth).

Perlman, D.L. and G. Adelson (1997) *Biodiversity: Exploring Values and Priorities in Conservation* (Malden: Blackwell).

Perring, M.P. and E.C. Ellis (2013) 'The Extent of Novel Ecosystems: Long in Time and Broad in Space' in R.J. Hobbs et al. (eds) *Novel Ecosystems: Intervening in the New Ecological World Order* (Chichester: Wiley-Blackwell).

Rassi, P., et al. (eds) (2010) *The 2010 Red List of Finnish Species* (Helsinki: Ministry of the Environment and Finnish Environment Institute).

Rawles, K. (2004) 'Biological Diversity and Conservation Policy' in M. Oksanen and J. Pietarinen (eds) *Philosophy and Biodiversity* (Cambridge: Cambridge University Press).

Regan, T. (1981) 'The Nature and the Possibility of an Environmental Ethic', *Environmental Ethics*, 3, 19–34.

Regan, T. (1988) *The Case for Animal Rights* (London: Routledge).

Sandler, R.L. (2010) 'The Value of Species and the Ethical Foundations of Assisted Colonization', *Conservation Biology*, 24, 424–31.

Sandler, R.L. (2012) *The Ethics of Species: An Introduction* (Cambridge: Cambridge University Press).
Sarkar, S. (1999) 'Wilderness Preservation and Biodiversity Conservation: Keeping Divergent Goals Distinct', *BioScience*, 49, 405–12.
Sarkar, S. (2002) 'Defining "Biodiversity", Assessing Biodiversity', *The Monist*, 85, 131–55.
Shapiro, J. (2001) *Mao's War against Nature* (Cambridge: Cambridge University Press).
Sharpe, V.A., et al. (eds) (2001) *Wolves and Human Communities: Biology, Politics, and Ethics* (Washington, DC: Island Press).
Svenning, J.-C. (2007) '"Pleistocene Re-Wilding" Merits Serious Consideration also Outside North America', *IBS Newsletter*, 5, 3–10.
Synthetic Biology (2013) http://syntheticbiology.org/FAQ.html (accessed 30 May 2013).
Taylor, P.W. (1986) *Respect for Nature: A Theory of Environmental Ethics* (Princeton: Princeton University Press).
Vuorisalo, T. and P. Laihonen (2000) 'Biodiversity Conservation in the North: History of Habitat and Species Protection in Finland', *Annales Zoologici Fennici*, 37, 281–97.
Westra, L. (1994) *An Environmental Proposal for Ethics: The Principle of Integrity* (Lanham: Rowman & Littlefield).
Wilson, E.O. (1994) *The Diversity of Life* (Harmondsworth: Penguin).
Zimov, S.A. (2005) 'Pleistocene Park: Return of the Mammoth's Ecosystem', *Science*, 308, 796–8.

Epilogue

Helena Siipi and Markku Oksanen

The aim of this book has not been to present a single unified view on resurrection science and its practical aspects. Rather the writers have been free to express their own views and formulate their own arguments over philosophical and ethical issues of de-extinction, rewilding and conservation gene-technology. In the end, nevertheless, it seems that the writers agree on many issues.

To begin with, it is evident that de-extinction and the gene-technological conservation is philosophically and biologically tricky. The idea of resurrection science rests on assumptions of the most fundamental kind in the philosophy of biology. It is impossible to take any stand for or against de-extinction without also taking a stand on those assumptions. Thus, by including de-extinction and genetic modification in one's conservation toolbox, one also ties oneself to certain views on species and extinction. Vice versa, holding certain views about species and extinction logically binds one to adopting certain views on the possibility and acceptability of de-extinction.

Most writers are willing to distinguish between two questions: the question of the best possible state of affairs and the question of appropriate actions in the current state of affairs. Probably all researchers working on conservation would prefer having a world without habitat fragmentation, climate change, pollution, biodiversity loss and other environmental damage. Most of them are also willing to agree that it would be prima facie better to aim at conservation goals and avoid the loss of species by non-interventionist *in situ* approaches than by relying on resurrection science. If this better world were attainable, it would be easy to reject emphatically any suggestions in favour of de-extinction and genetic modification. However, the world is not ideal and the non-interventionist methods have not fully carried out their promise in the face of biodiversity loss. In practice conservationists accept various *ex situ* methods in cases where a species is critically endangered. *Ex situ* methods include seed banks, botanical gardens and zoos, and these methods always involve intervention. In this situation, the option of de-extinction may be appealing. In the non-ideal world, the issue of de-extinction is far more complicated than in the ideal world where there would be no need for it.

None of the contributors argue for the categorical banning of de-extinction and genetic modification. On the other hand, none of the contributors are cheering for their unlimited application. It is, rather, cautiousness that characterizes the writers' views. They are definitely not ready to accept the view that de-extinction and gene-technological conservation are easy solutions that can (if and once the technology becomes advanced enough) solve the biodiversity crisis. Yet, there are probably cases where de-extinction and the gene-technological conservation are acceptable and even desirable. In other words, if a satisfactory level of biodiversity cannot be reached by the traditional methods, de-extinction and genetic modification could be the last resort. Saying this does not necessarily open a Pandora's box in conservation, as long as the precise characteristics of an emergency situation have been given.

This conclusion, even though being moderate, careful and far from surprising, leads to further questions. If de-extinction and the gene-technological conservation are in some cases possible and acceptable, how and on what conditions can they be implemented? Once implemented, how should they and the consequences of their use be controlled and monitored?

Another practical question concerns the selection of the criteria for suitable species and their habitats. Is it more acceptable, or do we have a stronger duty, to re-create species whose extinction is anthropogenic than otherwise vanished species? The passenger-pigeon is an example of a species that is considered to have been hunted to extinction by human beings (Zimmer 2013, pp. 36–7; Stone 2013, p. 19). The original enormity of the population in itself might, however, have resulted from human activity as well (Mann 2006, pp. 315–18; Herrmann and Woods 2010) and the volume of the population could be a symptom of ecological malfunction or 'illness'. The woolly mammoth case is far more nuanced: the drivers of its extinction may be multiple and vary between populations, which complicates any study. Is it, thus, ethically better to revive the passenger-pigeon than the woolly mammoth?

Not necessarily. The answer also depends on animal welfare factors as well as on the availability of suitable habitats for the animals. Ecosystems are constantly evolving and changing, which makes it difficult to identify suitable niches for the re-created species within their historic range. For example, the success and demise of the passenger-pigeon is closely linked with changes in land use and the disappearance of its major food source. In other cases, re-created species are adaptive and might produce a suitable habitat by their own means, like the woolly mammoth in Siberia (Pleistocene Park 2013) or the ancient breeds of cow and horse in the Oostwaardensplassen in the Netherlands. In this book, we have not discussed global climate change at great length. It is, however, crystal clear that predictions in regard to local changes must play a crucial role when decisions about the placements of de-extinct animals are made.

As noted in the chapters in this volume, the resurrection of extinct species raises concerns as to how they would affect ecosystems and individual animals. It also raises questions about human well-being, the freedom of science and political decision-making. As Rebecca J. Rosen (2012) puts it, 'if you could actually bring back anything, would you bring back the California grizzly bear? A species that could eat people?' Even if such direct threats were not involved, it is worth asking what the roles of stakeholders in decision-making regarding de-extinction and the gene-technological conservation of endangered species could and should be. If the Tasmanian tiger or a moa were released into a place near my home, should I have an opportunity to express my opinion? Apparently, some of those animals would end up living in laboratories, zoos or other confined facilities, which could be problematic. But what is their value and what is the value of having the species if it could only exist in isolation from nature? The technology, which will probably soon be available, has the potential to a considerable degree to change the environment around us – even to the magnitude of a Pleistocene rewilding. Thus, decisions come down to the question: what kind of environment will we have in the future? With such a fundamental question, how should the public be engaged in decision-making?

As always, scientific research and the development of applications require resources. De-extinction and gene-technological conservation are not going to be a bargain. Some researchers have already expressed their worries that there will be fewer resources for traditional conservation (Ehrenfeld 2006, p. 731; Rubenstein et al. 2006; Zimmer 2013, p. 40; Redford et al. 2013, p. 3; Stone 2013, p. 19). Sherkow and Greely (2013, p. 32) go as far as to suggest that de-extinction projects should not be publicly funded. This proposal is, however, unenforceable, because the scope of resurrection science can be understood in multiple ways. Therefore, it is impossible to demarcate amongst different lines of gene-technological development. Developing new techniques to improve the quality of ancient DNA may help to improve fertility treatments and vice versa. Although the withdrawal of public funding from the core projects of resurrection science could help to keep the present level of funding for conservation, the proposal has other drawbacks. It could leave – to some extent – the development of resurrection science to wealthy private venturers and their whims. Even though it is certainly 'cool' to re-create the charismatic woolly mammoth, re-creating the bucardo, the passenger-pigeon or gastric brooding frogs is less tempting. There is nothing bad in coolness or in charisma; nevertheless, they are unlikely to live up to conservational and scientific needs.

The commercial dimension of resurrection science is likely to focus on the production of animal companions or the clones of beloved dogs and cats. These animals would first and foremost be individuals to which their owners are attached. The owner, however, may well be disappointed by the outcome because the interplay between genomes and their environment may prevent the generation of exact copies. What would the limits for commercial use be? How and by whom should 'resurrection firms' be controlled?

The issue of de-extinction is one of the key subjects of what can be called the new conservation debate. It is a plainly interventionist and technologically advanced approach to conservation. As has been shown in the chapters in this volume, the technological solution offered by resurrection science has many problems, if not fatal flaws.

References

Ehrenfeld, D. (2006) 'Transgenics and Vertebrate Cloning as Tools for Species Conservation', *Conservation Biology*, 20(3), 723–32.

Herrmann, B. and W.I. Woods (2010) 'Neither Biblical Plague nor Pristine Myth: A Lesson from Central European Sparrows', *Geographical Review*, 100, 176–86.

Mann, M.C. (2006) *1491: The Americas Before Columbus* (London: Granta).

Pleistocene Park (2013) www.pleistocenepark.ru/en/ (accessed 30 August 2013).

Redford, K.H. et al. (2013) 'Synthetic Biology and Conservation of Nature: Wicked Problems and Wicked Solutions', *PLOS Biology*, 11(4), 1–4.

Rosen, R.J. (2012) 'Assuming We Develop the Capability, Should We Bring Back Extinct Species?', *The Atlantic*, 7 August.

Rubenstein, D.R. et al. (2006) 'Pleistocene Park: Does Re-wilding North America Represent Sound Conservation for the 21st Century?', *Biological Conservation*, 132, 232–8.

Sherkow, J.S. and H.T. Greely (2013) 'What If Extinction Is Not Forever?', *Science*, 340, 32–3.

Stone, R. (2013) 'Fluttering From the Ashes', *Science*, 340, 19.

Zimmer, C. (2013) 'Bringing Them Back to Life', *National Geographic*, April, 28–41.

Index

Printed and bound in the United States of America